T0212427

Lecture Notes in Computer Science   **11848**

More information about this series at http://www.springer.com/series/7412

Markus D. Schirmer · Archana Venkataraman ·
Islem Rekik · Minjeong Kim ·
Ai Wern Chung (Eds.)

Connectomics in NeuroImaging

Third International Workshop, CNI 2019
Held in Conjunction with MICCAI 2019
Shenzhen, China, October 13, 2019
Proceedings

 Springer

Editors
Markus D. Schirmer [iD]
Harvard Medical School
Boston, MA, USA

Archana Venkataraman [iD]
Johns Hopkins University
Baltimore, MD, USA

Islem Rekik
Istanbul Technical University
Istanbul, Turkey

Minjeong Kim
University of North Carolina
Greensboro, NC, USA

Ai Wern Chung [iD]
Harvard Medical School
Boston, MA, USA

ISSN 0302-9743   ISSN 1611-3349 (electronic)
Lecture Notes in Computer Science
ISBN 978-3-030-32390-5   ISBN 978-3-030-32391-2 (eBook)
https://doi.org/10.1007/978-3-030-32391-2

LNCS Sublibrary: SL6 – Image Processing, Computer Vision, Pattern Recognition, and Graphics

This Springer imprint is published by the registered company Springer Nature Switzerland AG
The registered company address is: Gewerbestrasse 11, 6330 Cham, Switzerland

Preface

The Third International Workshop on Connectomics in NeuroImaging (CNI 2019) was held in Shenzhen, China, on October 13, 2019, in conjunction with the 22nd International Conference on Medical Image Computing and Computer Assisted Intervention (MICCAI).

Connectomics is the study of whole brain association maps, i.e., the connectome, with a focus on understanding, quantifying, and visualizing brain network organization. Connectomics research is of interest to the neuroscientific community largely because of its potential to understand human cognition, its variation over development and aging, and its alteration in disease or injury. As such, big data in connectomics are rapidly growing with emerging international research initiatives collecting large, high-quality brain images with structural, diffusion, and functional imaging modalities. CNI aimed to propel research which leverages this increasing wealth of connectomic data. It brought together computational researchers (computer scientists, data scientists, computational neuroscientists) to discuss advancements in connectome construction, analysis, visualization, and their use in clinical diagnosis and group comparison studies. CNI 2019 was held as a single-track workshop that included two keynote speakers (Yong He from the Beijing Normal University, Beijing, China, and Fan Zhang, from Harvard Medical School, Boston, USA), oral paper presentations, and poster sessions.

Large, open source datasets, such as the Human Connectome Project (HCP) and the Autism Brain Imaging Data Exchange (ABIDE), have spurred the development of new and increasingly powerful machine learning strategies in connectomics, for which testing in a controlled setting is lacking. For the first time, CNI combined the workshop with a Transfer Learning Challenge. We provided training and validation sets of functional connectivity data of an attention deficit hyperactivity disorder (ADHD) cohort with age-matched neurotypical controls. The test data was withheld before the challenge to ensure comparability of the results. Therefore, CNI not only continued to showcase the latest contributions in this area, but acknowledged the challenge of validating new methodologies by providing a platform to address the open questions of their generalizability and clinical relevance.

The quality of submissions to our workshop was very high. Authors were asked to submit papers of 8–10 pages in length for review. A total of 14 papers were submitted to the workshop in response to our call for papers. Each of the 14 papers underwent a rigorous double-blind peer-review process, with each paper being reviewed by at least two reviewers from the Program Committee, composed of 20 well-known experts in the field of connectomics. Based on the reviewing scores and critiques, 13 papers were accepted for presentation at the workshop, and chosen to be included in this Springer LNCS volume. In order to allow the authors to address the reviews, the page limit was further extended per submission. The large variety of connectomics

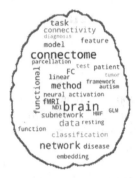

Fig. 1. Word cloud based on the abstracts of accepted submissions.

techniques applied in neuroimaging applications were well represented at the CNI 2019 workshop as demonstrated in Fig. 1.

We are grateful to the Steering and Program Committees for reviewing the submitted papers and giving constructive comments and critiques, to the authors for submitting high-quality papers, to the presenters for excellent presentations, and to all CNI 2019 attendees who came to Shenzhen from all around the world.

October 2019

Markus D. Schirmer
Archana Venkataraman
Islem Rekik
Minjeong Kim
Ai Wern Chung

Organization

General Chairs

Markus D. Schirmer   Harvard Medical School, USA
Archana Venkataraman   Johns Hopkins University, USA
Islem Rekik   Istanbul Technical University, Turkey
Minjeong Kim   University of North Carolina at Greensboro, USA
Ai Wern Chung   Harvard Medical School, USA

Steering Committee

Brent Munsell   University of North Carolina at Chapel Hill, USA
Guorong Wu   University of North Carolina at Chapel Hill, USA
Peipeng Liang   Capital Normal University, China

Program Committee

Gareth Ball   Murdoch Children's Research Institute, Australia
Dafnis Batalle   King's College London, UK
Brian Caffo   Johns Hopkins University, USA
Sheng He   Boston Children's Hospital, USA
Yoonmi Hong   University of North Carolina at Chapel Hill, USA
Kiho Im   Boston Children's Hospital, USA
Jaeil Kim   Kyungpook National University, South Korea
Peipeng Liang   Capital Normal University, China
Brent Munsell   University of North Carolina at Chapel Hill, USA
Mary Beth Nebel   Kennedy Krieger Institute, USA
Jonathan O'Muircheartaigh   King's College London, UK
Yangming Ou   Boston Children's Hospital, USA
Sanghyun Park   DGIST, South Korea
Mayssa Soussia   École Nationale d'Ingénieurs de Tunis, Tunisia
Heung-Il Suk   Korea University, South Korea
Matthew Toews   École de technologie supérieure, France
Guorong Wu   University of North Carolina, USA
Han Zhang   University of North Carolina at Chapel Hill, USA
Yu Zhang   Stanford University, USA
Lilla Zöllei   Massachusetts General Hospital, USA

Contents

Unsupervised Feature Selection via Adaptive Embedding and Sparse Learning for Parkinson's Disease Diagnosis

Zhongwei Huang[1], Haijun Lei[1], Guoliang Chen[1], Shiqi Li[1], Hancong Li[1], Ahmed Elazab[2], and Baiying Lei[2(✉)]

[1] Key Laboratory of Service Computing and Applications, Guangdong Province Key Laboratory of Popular High Performance Computers, College of Computer Science and Software Engineering, Shenzhen University, Shenzhen 518060, China
[2] National-Regional Key Technology Engineering Laboratory for Medical Ultrasound, Guangdong Key Laboratory for Biomedical Measurements and Ultrasound Imaging, School of Biomedical Engineering, Health Science Center, Shenzhen University, Shenzhen 518060, China
leiby@szu.edu.cn

Abstract. Parkinson's disease (PD) is known as a progressive neurodegenerative disease in elderly people. Apart from decelerating the disease exacerbation, early and accurate diagnosis also alleviates mental and physical sufferings and provides timely and appropriate medication. In this paper, we propose an unsupervised feature selection method via adaptive manifold embedding and sparse learning exploiting longitudinal multimodal neuroimaging data for classification and regression prediction. Specifically, the proposed method simultaneously carries out feature selection and dynamic local structure learning to obtain the structural information inherent in the neuroimaging data. We conduct extensive experiments on the publicly available Parkinson's progression markers initiative (PPMI) dataset to validate the proposed method. Our proposed method outperforms other state-of-the-art methods in terms of classification and regression prediction performance.

Keywords: Parkinson's disease · Unsupervised feature selection · Adaptive manifold embedding · Classification · Regression prediction

1 Introduction

Parkinson's disease (PD) is a progressive neurodegenerative disorder in the elderly people. The symptoms of PD progressively occur and continue to be worsen, thus middle or late patients usually have enduring mental and physical torment and even life threatening conditions. PD mainly has four symptoms: muscle rigidity, bradykinesia, rest tremor, and postural instability. Apart from

M. D. Schirmer et al. (Eds.): CNI 2019, LNCS 11848, pp. 1–9, 2019.
https://doi.org/10.1007/978-3-030-32391-2_1

these external symptoms, there are also accompanying symptoms such as depression, sleep, olfaction, and cognition disturbances [4]. These symptoms appear primarily due to the degeneration of dopaminergic neurons in the substantianigra [7]. In the meantime, no dopaminergic deficiency was observed in some PD sufferers, namely, scans without evidence of dopamine deficit (SWEDD). This also raises the difficulty of PD diagnosis. Hence accurate PD diagnosis is essential, which can decelerate the disease exacerbation further and abate the physical and mental sufferings of patients.

Since multimodal data can supply complementary information for computer-aided PD diagnosis, multimodal data has attracted much attention and increasingly played a vital role in this area of work [12]. However, the limited number of subjects and multi-modal neuroimaging data usually has high feature dimensionality, which may cause overfitting issue and render the generalization model quite difficult. Although deep learning has been widely used in medical image analysis and achieved good performance, it is hard to build a good and robust model using a small number of subjects [6]. To overcome this weakness, feature selection is an impactful way using either supervised or unsupervised method by discovering disease-related characteristics [9]. Most supervised methods are based on single-task [13] or multi-task models [5]. The latter generally has better performance. However, there are two problems with multi-task methods. On the one hand, these methods forces to build a linear relationship of data to multi-task targets but they lose sight of the learning of the structural information inherent in data. On the other hand, these supervised methods need to provide additional labels and scores information. By contrast, unsupervised feature selection methods emphasize more on learning the structural information inherent in data. Most of them are based on filter methods [2], or embedded methods [10]. The latter generally has better performance and thus receives wide-ranging attention. However, there are two problems with embedding methods. First, they conduct local structure learning and sparse regression, respectively. Second, when learning local manifold structure, the similarity matrix obtained by the conventional embedding methods usually does not have a suitable neighborhood assignment. In other words, the connected components of the ideal similarity matrix should be the same as the number of classification. In the meantime, most existing studies mainly exploited baseline data to perform classification or clinical scores prediction. The longitudinal data (with multi-time points) is often ignored. However, due to the irreversible and sustained deterioration of the disease, it is important to build a good and robust diagnostic model for longitudinal data.

Inspired by the above, we propose an unsupervised feature selection method via adaptive manifold embedding and sparse learning based on longitudinal multimodal data. Overall, we show the main contributions of this study as follows:

1. We propose an unsupervised method that jointly carries out feature selection and dynamic local structure learning to obtain discriminative features.
2. We make the connected numbers of the similarity matrix equal to the number of classifications to obtain the inherent structural information of the data.

3. We conduct extensive experiments to validate the effectiveness of our method on PPMI dataset. Particularly, we exploit longitudinal data to increase the class label identification performance and achieve high accuracy compared with the state-of-art methods.

2   Methodology

2.1   System Overview

The whole procedures for classification and clinical scores prediction are shown in Fig. 1. First, we respectively extract 116-dimensional features from gray matter (GM) from magnetic resonance imaging (MRI), first eigenvalue (L1) and first eigenvector (V1) from diffusive tensor imaging (DTI). Then, we linearly combine these features and perform feature selection via adaptive manifold embedding and sparse learning. Finally, we exploit support vector machine to train a classification model and four regression models in baseline multimodal data. Specifically, we propose a new objective function to select discriminative features in baseline multimodal data and then build a classification model for classification in the baseline, 12 months, and 24 months data, respectively. Also, we build four regression models for predicting the clinical scores in baseline multimodal data.

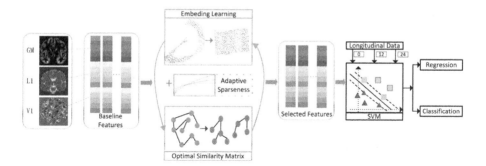

Fig. 1. The frame of the proposed method via adaptive manifold embedding and sparse learning.

2.2   Notation

In this paper, we use uppercase bold letters (e.g., $\mathbf{A}$) as matrices and lowercase bold characters (e.g., $\mathbf{a}$) as vectors. For a matrix $\mathbf{A} = [a_{kj}]$, $\mathbf{a}^k$ denotes the k-th row of $\mathbf{A}$ and $tr(\mathbf{A})$ denotes the trade. $\mathbf{A}^T$ denotes the transpose of $\mathbf{A}$. We denote $l_{2,p}$ norm of $\mathbf{A}$ as $||\mathbf{A}||_{2,p} = (\sum_k ||\mathbf{A}||_2^p)^{\frac{1}{p}}$.

2.3   Proposed Method

Let $\mathbf{A} \in \mathbf{R}^{n \times d}$ and $\mathbf{S} \in \mathbf{R}^{n \times n}$ represent the original high-dimensional data and the similarity matrix of n subjects and d features, respectively, where $\mathbf{a}^k$ denote k-th sample of $\mathbf{A}$, s_{kj} is a value of $\mathbf{S}$. Generally, we calculate $\mathbf{S}$ by the following function:

$$\min \sum_{k,j} (||\mathbf{a}^k - \mathbf{a}^j||_2^2 s_{kj} + \mu s_{kj}^2), \qquad s.t. \mathbf{s}^k \mathbf{1} = \mathbf{1}, 0 \le s_{kj} \le 1, \qquad (1)$$

where μ is a regularization parameter to avoid useless solutions. The similarity matrix obtained by Eq. 1 usually does not have a suitable neighborhood assignment. In other words, the connected components of the ideal similarity matrix should be the same as the number of classification (e.g., r). However, it is almost impossible to achieve the above requirement using Eq. 1. To solve the problem, we can make the rank of Laplacian matrix $\mathbf{L}$ of $\mathbf{S}$ equal to $n-r$, namely, $rank(\mathbf{L}) = n - r$. By this way, the similarity matrix will contain r connected components [8]. We add this constraint to Eq. 1 and then we have:

$$\min \sum_{k,j} (||\mathbf{a}^k - \mathbf{a}^j||_2^2 s_{kj} + \mu s_{kj}^2), \qquad s.t. \mathbf{s}^k \mathbf{1} = \mathbf{1}, 0 \le s_{kj} \le 1, rank(\mathbf{L}) = n - r,$$
$$(2)$$

where $\mathbf{L} = \mathbf{D} - \frac{\mathbf{S}^T + \mathbf{S}}{2}$, $\mathbf{D}$ is a diagonal matrix whose k-th diagonal value is $\sum_j \frac{s_{kj} + s_{jk}}{2}$. Since $rank(\mathbf{L}) = n - r$ also depends on the similarity matrix $\mathbf{S}$, it is hard to optimize Eq. 2. To solve it, let $\Psi_k(L)$ denote the k-th smallest eigenvalue of $\mathbf{L}$. Owing to positive semi-definiteness of $\mathbf{L}$, we easily get $\Psi_k(\mathbf{L}) \ge 0$. In the meantime, it can be easily known that $rank(\mathbf{L}) = n - r$ denotes $\sum_{k=1}^{r} \Psi_k(\mathbf{L}) = 0$. Because the derivation of $\sum_{k=1}^{r} \Psi_k(\mathbf{L})$ is difficult to solve, we use Ky Fan's Theorem [1] to obtain:

$$\sum_{k=1}^{r} \Psi_k(\mathbf{L}) = \min Tr(\mathbf{Q}^T \mathbf{L} \mathbf{Q}), \qquad s.t. \quad \mathbf{Q} \in \mathbf{R}^{n \times r}, \mathbf{Q}^T \mathbf{Q} = \mathbf{I}. \qquad (3)$$

Further, we can rewrite Eq. 2 as follows:

$$\min \sum_{k,j} (||\mathbf{a}^k - \mathbf{a}^j||_2^2 s_{kj} + \mu s_{kj}^2) + \sigma tr(\mathbf{Q}^T \mathbf{L} \mathbf{Q}),$$
$$s.t. \quad \mathbf{s}^k \mathbf{1} = \mathbf{1}, 0 \le s_{kj} \le 1, \mathbf{Q} \in \mathbf{R}^{n \times r}, \mathbf{Q}^T \mathbf{Q} = \mathbf{I}, \qquad (4)$$

where σ is a parameter which can be automatically adjusted in each iteration to make the connected component of $\mathbf{S}$ equal to r. In Eq. 4, the similarity matrix S is computed with the original high-dimensional feature space. However, multi-modal data often has many noisy and redundant features. To solve this problem and gain the sparse solution of the original feature, we jointly perform feature selection and adaptive manifold learning as follows:

$$\min \sum_{k,j}(||\mathbf{a}^k - \mathbf{a}^j||_2^2 s_{kj} + \mu s_{kj}^2) + \sigma tr(\mathbf{Q}^T \mathbf{L}\mathbf{Q}) + \lambda||\mathbf{W}||_{2,p}^p,$$

$$s.t. \quad \mathbf{s}^k \mathbf{1} = 1, 0 \le s_{kj} \le 1, \mathbf{Q} \in \mathbf{R}^{n \times r}, \mathbf{Q}^T \mathbf{Q} = \mathbf{I}, \mathbf{W}^T \mathbf{W} = \mathbf{I}, \tag{5}$$

where $\mathbf{W} \in \mathbf{R}^{d \times c}$ represents the weight coefficient of features, d and c are the original dimension and projection dimension, respectively. λ denotes weighting parameter, and the larger its value, the fewer features are selected. Meanwhile, we exploit multiple regularizers, namely, $l_{2,p}$ norm, to conduct adaptive sparse control for obtaining the most discriminative features according to different cases. Since the high-dimensional data easily makes the covariance matrix of $\mathbf{A}$ become singular, we introduce the constraint $\mathbf{W}^T \mathbf{W} = \mathbf{I}$ to obtain discriminative features. Finally, Fig. 2 presents the algorithm to solve Eq. 5 and its convergence. Due to the space finite, the derivation of this algorithm will be provided in its extended journal paper.

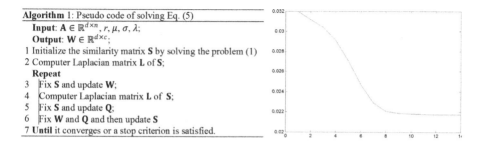

Fig. 2. The algorithm to solve Eq. 5 and its convergence.

3 Experiments

In this study, we use PPMI dataset for performance evaluation. We collected baseline data obtained from 238 samples including 62 normal control (NC), 142 PD, and 34 SWEDD samples. We also collected 12 months data obtained from 186 including 54 NC, 123 PD, and 9 SWEDD samples and collected 24 months data obtained from 127 samples including 7 NC, 98 PD, and 22 SWEDD samples. Meanwhile, we used Geriatric Depression Scale, Epworth Sleepiness Scale, University of Pennsylvania Smell Identification Test, and Montreal Cognitive Assessment (MoCA) to evaluate depression, sleep, olfaction, and cognition scores, respectively.

3.1 Image Preprocessing

For MRI data preprocessing, we first conduct anterior commissure-posterior commissure (AC-PC) reorientation and then exploit the voxel-based morphometric tool [11] to segment MRI images for obtaining GM tissue. Further, we register the GM with the automated anatomical labeling (AAL) atlas to obtain

116-dimensional features. For DTI data preprocessing, we use the FMRIB software library toolbox [3] to correct eddy current distortion and further compute L1 and V1 data. Finally, we conduct AC-PC reorientation on L1 and V1 data and then use AAL atlas to obtain 116-dimensional features from L1 and V1 data, respectively.

3.2 Experimental Setting

In the paper, on the whole pipeline, we use a 10-fold cross-validation method to verify the effectiveness of our method in baseline data. Specifically, we perform two classification tasks (i.e., NC vs. PD and NC vs. SWEDD) and four scores prediction (i.e., depression, sleep, olfaction, cognition scores) in baseline data. To verify the robustness and accuracy of the proposed method on longitudinal time, we also use 12 months and 24 months data as test dataset to validate classification performance. We use quantitative measurements to evaluate classification performance, namely, accuracy (ACC), sensitivity (SEN), precision (PREC), and area under the receiver operating characteristic (ROC) curve (AUC). To estimate regression performance, we compute the Pearson's correlation coefficient (CC) and root mean squared error (RMSE) between the predicted and actual clinical scores.

We also compare the proposed method with state-of-the-art methods including: (1) The Laplacian score (Lscore) method for unsupervised feature selection [2]; (2) The robust spectral feature selection (RSFS) method simultaneously using flexible manifold embedding and l_1 norm for robust unsupervised feature selection [10]; (3) The multimodal multi-task (M3T) method exploiting $l_{2,1}$ norm to gain a common feature subset of multi-task for supervised feature selection [12]; (4) The multimodal sparse learning (MMSL) method considering the relations among rows and columns in response matrices for supervised feature selection [5].

3.3 Classification Performance

Table 1 presents the classification performances in longitudinal multimodal data. We can see that the proposed method has the best classification performance. Meanwhile, in the longitudinal direction, our method achieves the most robust performance, such as the accuracies of 81.45%, 80.23%, and 97.14% in NC vs. PD and the accuracies of 89.56%, 95.24%, and 82.76% in NC vs. SEDDD, respectively.

Traditionally, unsupervised methods are more difficult than the supervised one for the absence of label information. However, our method exhibits better classification performance than M3T and MMSL methods. For example, the proposed method has higher accuracies than the MMSL method in baseline data, such as 81.45% vs. 81.33% for NC and PD and 89.56% vs. 87.44% for NC and SWEDD. The reason is that our method can effectively capture the structural information inherent in data. In addition, Fig. 3 also shows the ROC curves for different methods, which further presents the good performance of

Table 1. Classification performances for longitudinal data in all methods.

Time	Method	NC vs. PD				NC vs. SWEDD			
		ACC	SEN	PREC	AUC	ACC	SEN	PREC	AUC
0 min	Lscore	79.45	63.57	70.60	74.36	83.67	98.33	81.41	75.32
	RSFS	79.43	**66.67**	69.94	74.41	85.56	100.00	82.79	74.70
	M3T	78.90	62.62	68.88	72.62	82.44	96.67	80.98	68.83
	MMSL	81.33	64.52	76.06	**75.28**	87.44	**100.00**	85.24	76.71
	Proposed	**81.45**	64.76	**76.90**	73.59	**89.56**	98.33	**87.62**	**81.85**
12 min	Lscore	80.23	62.96	69.39	77.36	90.48	100.00	90.00	84.16
	RSFS	**81.92**	68.52	71.15	80.13	92.06	96.30	94.55	85.80
	M3T	80.23	64.81	68.63	79.01	90.48	96.30	92.86	81.07
	MMSL	81.36	64.81	**71.43**	78.65	92.06	96.30	94.55	84.36
	Proposed	80.23	**81.48**	63.77	**81.21**	95.24	100.00	94.74	**87.24**
24 min	Lscore	96.19	57.14	80.00	79.15	58.62	85.71	35.29	74.03
	RSFS	97.14	71.43	83.33	80.03	72.41	100.00	46.67	**92.86**
	M3T	93.33	71.43	50.00	88.19	65.52	100.00	41.18	92.86
	MMSL	96.19	71.43	71.43	**93.15**	72.41	100.00	46.67	88.31
	Proposed	**97.14**	**71.43**	**83.33**	88.05	82.76	100.00	58.33	89.61

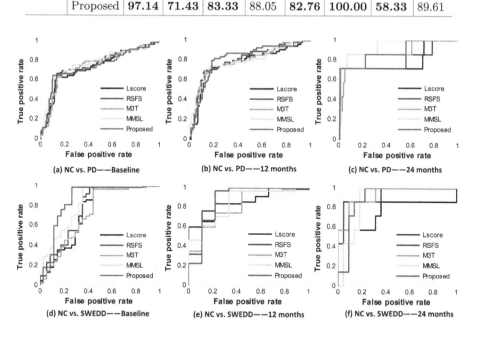

Fig. 3. ROC curves for the longitudinal multimodal data in all methods.

our method. Finally, Fig. 4 shows top brain regions that contribute most to the learned common structure and their connection network, which can help researchers to further study in the future.

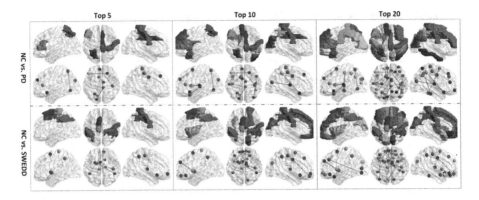

Fig. 4. Top brain regions that contribute most to the learned common structure and their connection network.

3.4 Regression Performance

We use CC and RMSE to estimate the regression performance in baseline data. In NC vs. PD, the MMSL achieves the best performance for predicting depression and sleep scores. The best performance using CC and RMSE is 0.5699 and 4.4037 in depression score, and 0.5694 and 5.7426 in sleep score. Our method has best performance in the prediction of olfaction and cognition scores. The best performance is 0.5526 (CC) and 8.3057 (RMSE) in olfaction score, and 0.6046 (CC) and 4.5603 (RMSE) in cognition score. In NC vs. SWEDD, our method achieves the best performance for predicting depression, sleep, olfaction, and cognition scores. The best performance using CC and RMSE is 0.7727 and 3.1047, 0.7583 and 4.3512, 0.7932 and 5.2154, and 0.7998 and 2.8487 in the four scores, respectively. Figure 5 also shows regression performance of the competing methods. We can see that our method has the best performance overall.

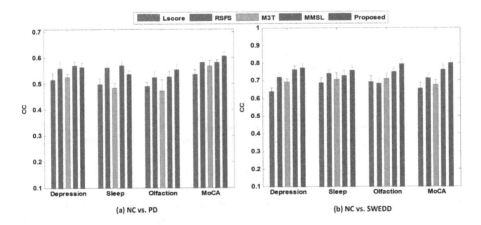

Fig. 5. Regression performance of the competing methods in baseline data.

4 Conclusion

In this paper, an unsupervised feature selection method is proposed to simultaneously carry out adaptive sparse learning by exploiting $l_{2,p}$ norm and local structure learning to select discriminative features. Extensive experiments were performed to validate the effectiveness of the proposed method on PPMI dataset. The longitudinal experimental results demonstrated that our method can strengthen the performance in class label identification and outperforms the state-of-art methods as well. Meanwhile, our unsupervised method is superior to other unsupervised methods in clinical scores prediction and has the best regression performance.

Acknowledgments. This work was supported partly by the Integration Project of Production Teaching and Research by Guangdong Province and Ministry of Education (No. 2012B091100495), Shenzhen Key Basic Research Project (No. JCYJ20170302153337765), Guangdong Pre-national Project (No. 2014GKXM054), and Guangdong Province Key Laboratory of Popular High Performance Computers (No. 2017B030314073).

References

1. Fan, K.: On a theorem of Weyl concerning eigenvalues of linear transformations I. Proc. Nat. Acad. Sci. U.S.A. **35**(11), 652 (1949)
2. He, X., Cai, D., Niyogi, P.: Laplacian score for feature selection. In: Advances in Neural Information Processing Systems, pp. 507–514 (2006)
3. Jenkinson, M., Beckmann, C.F., Behrens, T.E., Woolrich, M.W., Smith, S.M.: FSL. Neuroimage **62**(2), 782–790 (2012)
4. Kalia, L.V., Lang, A.E.: Parkinson's disease. Lancet (Lond. Engl.) **386**(9996), 896–912 (2015). https://doi.org/10.1016/S0140-6736(14)61393-3
5. Lei, H., et al.: Joint detection and clinical score prediction in Parkinson's disease via multi-modal sparse learning. Expert Syst. Appl. **80**, 284–296 (2017)
6. Litjens, G., et al.: A survey on deep learning in medical image analysis. Med. Image Anal. **42**, 60–88 (2017)
7. Lotharius, J., Brundin, P.: Pathogenesis of Parkinson's disease: dopamine, vesicles and -synuclein. Nat. Rev. Neurosci. **3**(12), 932 (2002)
8. Mohar, B., Alavi, Y., Chartrand, G., Oellermann, O.R.: The Laplacian spectrum of graphs. Graph Theory. Comb. Appl. **2**(871–898), 12 (1991)
9. Nie, F., Zhu, W., Li, X.: Unsupervised feature selection with structured graph optimization. In: Thirtieth AAAI Conference on Artificial Intelligence (2016)
10. Shi, L., Du, L., Shen, Y.D.: Robust spectral learning for unsupervised feature selection. In: 2014 IEEE International Conference on Data Mining, pp. 977–982. IEEE (2014)
11. Whitwell, J.L.: Voxel-based morphometry: an automated technique for assessing structural changes in the brain. J. Neurosci. **29**(31), 9661–9664 (2009)
12. Zhang, D., Shen, D., Initiative, A.D.N.: Multi-modal multi-task learning for joint prediction of multiple regression and classification variables in Alzheimer's disease. NeuroImage **59**(2), 895–907 (2012)
13. Zou, H., Hastie, T.: Regularization and variable selection via the elastic net. J. R. Stat. Soc.: Ser. B (Stat. Methodol.) **67**(2), 301–320 (2005)

A Novel Graph Neural Network to Localize Eloquent Cortex in Brain Tumor Patients from Resting-State fMRI Connectivity

Naresh Nandakumar[1](✉), Komal Manzoor[2], Jay J. Pillai[2], Sachin K. Gujar[2], Haris I. Sair[2], and Archana Venkataraman[1]

[1] Department of Electrical and Computer Engineering,
Johns Hopkins University, Baltimore, USA
nnandak1@jhu.edu
[2] Department of Neuroradiology,
Johns Hopkins School of Medicine, Baltimore, USA

Abstract. We develop a novel method to localize the language and motor areas of the eloquent cortex in brain tumor patients based on resting-state fMRI (rs-fMRI) connectivity. Our method leverages the representation power of convolutional neural networks through specialized filters that act topologically on the rs-fMRI connectivity data. This Graph Neural Network (GNN) classifies each parcel in the brain into eloquent cortex, tumor, or background gray matter, thus accommodating varying tumor characteristics across patients. Our loss function also reflects the large class-imbalance present in our data. We evaluate our GNN on rs-fMRI data from 60 brain tumor patients with different tumor sizes and locations. We use motor and language task fMRI for validation. Our model achieves better localization than linear SVM, random forest, and a multilayer perceptron architecture. Our GNN is able to correctly identify bilateral language areas in the brain even when trained on patients whose language network is lateralized to the left hemisphere.

Keywords: Rs-fMRI · Graph Neural Network · Language localization

1 Introduction

The eloquent cortex consists of sensorimotor and language areas in the brain that are essential for human functioning. Given its importance, localizing and subsequently avoiding the eloquent cortex is a crucial step when planning a neurosurgery. However, this localization is challenging due to the varying anatomical boundaries of these networks and the effects of the tumor. For example, it has been shown that motor and language functionality in brain tumor patients can be displaced due to neural plasticity [1]. The gold standard for eloquent mapping is intraoperative electrical stimulation, which is highly invasive and requires the

© Springer Nature Switzerland AG 2019
M. D. Schirmer et al. (Eds.): CNI 2019, LNCS 11848, pp. 10–20, 2019.
https://doi.org/10.1007/978-3-030-32391-2_2

patient to be awake during surgery. The noninvasive alternative is task-fMRI. However, severely impaired patients, such as those with advanced brain tumors, may not be able to perform these tasks, thus reducing the reliability of the fMRI activation maps. Resting-state fMRI (rs-fMRI) captures spontaneous fluctuations in the brain, which can be used to identify functional systems in the absence of an experimental paradigm. Hence, rs-fMRI may provide an alternative for motor and language localization in critically ill patients [2].

Automatically identifying the eloquent cortex in brain tumor patients is a challenging problem with limited success in the literature. The work of [3] addresses the problem of shifting anatomical boundaries by matching functional brain regions across individuals via a diffusion map representation of task-fMRI. However, this method has yet to generalize to rs-fMRI. With regards to rs-fMRI, the work of [4] describes a method to obtain subject-specific functional parcellations of brain tumor patients using a Markov Random Field prior. However, this method is validated on a coarse functional parcellation which is unsuitible for presurgical mapping. The work of [5] describes a method to compute language laterality from rs-fMRI by comparing connectivity between fixed areas of expected language activation. However, this study stopped short of *localization*, which is the main clinical need. The authors of [2] propose a semi-automated method to determine the language network from group ICA maps of rs-fMRI data. However, this method relies on manual thresholding for each patient. Finally, the work of [6] describes a multi-layer perceptron to classify resting-state networks at the voxel level based on seed correlation maps, which was then extended to identify the language network in three separate tumor cases [7]. However, this method is computationally expensive, requires a large amount of training data, and has only been evaluated on a limited dataset.

In this paper, we propose the first end-to-end model that uses convolutional neural networks (CNNs) to identify eloquent cortex in brain tumor patients. Our problem loosely resembles image segmentation, for which deep learning approaches using CNNs have made great strides [8]. However, rs-fMRI captures correlated patterns of activity rather than local similarities, which cannot be represented by a traditional spatial convolution. Therefore, deep learning for rs-fMRI has focused almost exclusively on perceptron architectures [6] and patient wise classification [9], rather than network analysis. Our approach blends the ideas of image segmentation and functional network extraction. Namely, we construct a similarity graph from rs-fMRI data that summarizes functional connectivity between ROIs. These graphs are then input to a novel graph neural network (GNN) which leverages convolutional filters designed to act topologically upon similarity matrices [10]. The output of our GNN is a vector that classifies each node in the graph as either eloquent cortex, tumor, or background gray matter. We train and evaluate four separate GNN's to perform either language or motor classification. The motor classes are divided into three regions of the motor strip corresponding to finger, tongue, or foot movements. Our loss function reflects the large class-imbalance in our data, as eloquent cortex and tumor represent a

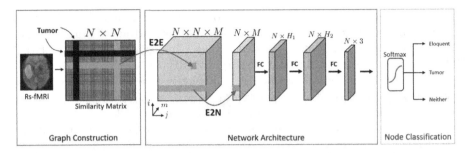

Fig. 1. The overall workflow of our model. Left: Graph construction encodes fMRI and tumor information. Middle: Our GNN architecture employs E2E, E2N, and FC layers for feature extraction. Right: We perform a node (parcel) identification task.

small fraction of the brain. Our model outperforms three baseline approaches in eloquent cortex detection, overall accuracy, and AUC.

2   A Graph Neural Network for Node Identification

The underlying assumption of our framework is that, while the anatomical boundaries of the eloquent cortex, particularly the language network, may shift, its connectivity with the rest of the brain will remain consistent [2]. We construct a weighted graph from the rs-fMRI data. We use a deep learning framework to capture complex interactions in the connectivity data. Our GNN node classifier extracts salient edge-node relationships and node features within the graph using a combination of specialized convolutional filters and fully-connected (FC) layers. An important distinction in our problem is the presence of large anatomical lesions, i.e, the brain tumors. Since the tumors often encroach into the gray matter, we introduce "missing" full rows and columns into our graph. These missing rows and columns are the most salient features of the data, therfore we introduce two baseline class labels, "tumor" and "background gray matter" to avoid biasing the algorithm. Figure 1 outlines our overall pipeline from graph construction to node classification.

Graph Construction. Let N be the number of brain regions in our parcellation and T be the number of time points for a rs-fMRI scan. We define $\mathbf{x}_i \in \mathbb{R}^{T \times 1}$ as the average time series extracted from parcel i. We construct graph $\mathbf{W} \in \mathbb{R}^{N \times N}$

$$\mathbf{W}_{i,j} = \exp\left[\frac{\langle \mathbf{x}_i, \mathbf{x}_j \rangle}{\epsilon} - 1\right] \tag{1}$$

where $\langle \cdot, \cdot \rangle$ represents the Pearson's correlation coefficient between time courses and $\epsilon \geq 1$ controls decay speed. By construction, rows and columns of $\mathbf{W}$ that correspond to tumor are "missing" and computationally set to zero to indicate that they are not functionally similar to any other region in the brain. Our choice of $\epsilon \geq 1$ along with the form of Eq. (1) asserts that $\mathbf{W}_{i,j} > 0$ for all non-tumor regions. Therefore, even two healthy parcels with a strong negative correlation will still be more functionally similar than tumor regions in our model. Our framework assumes that tumor boundaries have been delineated for each patient.

Neural Network Architecture. Our GNN architecture employs both convolutional and FC layers to process node information. While traditional convolutional layers assume a grid-like organization to extract spatially local features, our GNN uses one edge-to-egde (E2E) and one edge-to-node (E2N) layer developed in [10], which act topologically on similarity graph data. These convolutional filters span full rows and columns of the graph and were originally designed to perform regression from diffusion MRI connectivity. Mathematically, an E2E filter is composed one row filter, one column filter, and a learned bias, which totals $2N + 1$ parameters. Let $m \in \{1, \cdots, M\}$ be the E2E filter index, $\mathbf{r}^m \in \mathbb{R}^{1 \times N}$ be the m-th row filter, $\mathbf{c}^m \in \mathbb{R}^{N \times 1}$ be the m-th column filter and $\mathbf{b} \in \mathbb{R}^{M \times 1}$ be the E2E bias. The feature map $\mathbf{A}^m \in \mathbb{R}^{N \times N}$ output from E2E filter m and activation function ϕ is computed as

$$\mathbf{A}_{i,j}^m = \phi\Big(\sum_{n=1}^{N} \mathbf{r}_n^m \mathbf{W}_{i,n} + \mathbf{c}_n^m \mathbf{W}_{n,j} + \mathbf{b}_m \Big) \tag{2}$$

Intuitively, an E2E filter for node pair (i, j) computes a weighted sum of edge strengths over all edges connected to either node i or j. Even with symmetric input $\mathbf{W}$, E2E filters and corresponding feature maps are not necessarily symmetric. This asymmetry is desirable, as functional systems in the brain tend to be lateralized. We use the E2E layer to encode multiple different views (maps) of the edge-to-egde similarities within our connectome data.

The E2N layer condenses our representation from size $N \times N \times M$ after the E2E layer to $N \times M$, analogous to M features for each node. The E2N filter is simply a 1D convolution along the columns of each feature map. Let $\mathbf{g}^m \in \mathbb{R}^{N \times 1}$ be the m-th E2N filter and $\mathbf{d} \in \mathbb{R}^{M \times 1}$ be the E2N bias. The E2N output $\mathbf{a}^m \in \mathbb{R}^{N \times 1}$ from input $\mathbf{A}^m$ is computed as

$$\mathbf{a}_i^m = \phi\Big(\sum_{n=1}^{N} \mathbf{g}_n^m \mathbf{A}_{n,i}^m + \mathbf{d}_m \Big). \tag{3}$$

Mathematically, the E2N filter computes a single value for each node i by taking a weighted combination of edges associated with it. Our motivation for using this layer is to collapse our representation along the second dimension to obtain M features for each node. This step is similar in nature to extracting graph theoretic features, such as node centrality. In particular, we have a representation that encodes the relationship each node has to its connectivity matrix [11].

Our node identification network uses a cascade of three FC layers of sizes $M \times H_1$, $H_1 \times H_2$ and $H_2 \times 3$ respectively. We apply activation functions between each layer. The FC layers find nonlinear combinations of the features to best discriminate class membership for each brain parcel. Overall, our network takes $N \times N$ input and outputs an $N \times 3$ matrix for classification. Notice that the first input dimension N is maintained throughout our whole network and is not transformed. Therefore, our network maintains node structure to ultimately discriminate class membership for all nodes within one connectome at a time. As shown in Fig. 1, one design choice we make is to set $H_2 > H_1$. Empircally, this relationship robustly captures the structure of our class membership.

Weighted Loss Function. Naturally, there exists a large class imbalance in our setup, as the majority of nodes considered will be background gray matter. We cannot rely on traditional data augmentation techniques to mitigate this imbalance, as our model operates on whole-brain connectivity. To accomodate for the class imbalance, we train our model with a modified version of the Risk-sensitive cross-entropy (RSCE) loss function [12], which is designed to handle membership imbalance in multi-class classification. Let $\hat{y}_c^n$ be the output probability of our network for assigning node n to class c and y_c^n be 1 when node n belongs to class c and 0 otherwise. The loss function per patient is

$$\mathcal{L}(y_c^n, \hat{y}_c^n) = -\frac{1}{N} \sum_{n=1}^{N} \sum_{c=1}^{C} \delta_c \cdot y_c^n \log(\hat{y}_c^n) \qquad (4)$$

where δ_c is the risk factor associated with class c. If δ_c is small, then we pay a smaller penalty for misclassifying samples that belong to class c. Our strategy is to penalize misclassifying eloquent nodes (false negatives) larger than misclassifying background (false positives) to encourage our model to learn the language and motor distributions given a small number of language training samples.

Neural Network Implementation Details. We implement our network in PyTorch using the SGD optimizer with weight decay $= 5 \times 10^{-5}$ for parameter stability, and momentum $= 0.9$ to improve convergence. For our model, $\epsilon = 1$ and layer dimensions are $M = 8$, $H_1 = 9$, $H_2 = 27$ and $C = 3$. We train our model with learning rate .005 and 80 epochs, which provides for reliable performance without overfitting. The LeakyReLU$(x) = \max(0, x) + 0.33 \cdot \min(0, x)$ activation function is applied at each hidden layer. Empirically, this activation function is robust to a range of initializations. A softmax activation is applied at the final layer for classification. After cross-validation, we set $\delta = (1.3, .3, .15)$ for the eloquent cortex, tumor, and background gray matter classes respectively. With GPU, total training time is within 3–5 min.

2.1 Baseline Comparisons

We evaluate the performance of our GNN against 3 baseline methods. The first baseline is a linear SVM based on graph theoretic measures: node degree, betweenness, closeness, and eigenvector centrality [11]. The second baseline is a random forest (RF) on the stacked rs-fMRI similarity features of each node. We omit tumor class and nodes for SVM and RF as the algorithms do not exploit the spatial consistency of the similarity matrix. The last baseline is a multi-layer perceptron (MLP) to observe how adding specialized E2E and E2N layers changes performance for this task. The MLP maintains the same input-output relationship, total parameter number, activations, and loss function as the GNN.

3   Experimental Results

Dataset and Preprocessing: We evaluate the GNN on rs-fMRI data from 60 patients who underwent preoperative mapping as part of their presurgical workup. The data was acquired using a 3.0 T Siemens Trio Tim (TR = 2000 ms, TE = 30 ms, FOV = 24 cm, res = $3.59 \times 3.59 \times 5$ mm). The fMRI was processed using SPM8. The steps include slice timing correction, motion correction and registration to the MNI-152 template. The rs-fMRI was bandpass filtered from 0.01 to 0.1 Hz, spatially smoothed with a 6 mm FWHM Gaussian kernel, scrubbed using the ArtRepair toolbox in SPM, linearly detrended, and underwent nuisance regression using CompCor.

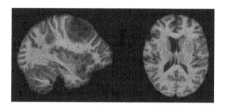

Fig. 2. Left: One left-hemisphere language network (red) subject. Right: One bilateral language network subject. (Color figure online)

Our dataset includes three different motor paradigms that were designed to target distinct parts of the motor homonculus [13]: finger tapping, tongue moving, and foot tapping. Since the task-fMRI data was acquired for clinical purposes, only 38 patients performed the finger task, 41 patients performed the tongue task, and 18 patients performed the foot task. Our ground truth language annotations are derived from task-fMRI activations of the same 60 patients during two language paradigms: sentence completion (SC) and silent word generation (SWG). Our dataset includes 55 patients with left-hemisphere language networks and 5 patients with bilateral networks. The fMRI underwent slice timing correction, motion correction and registration to the MNI-152 template. The General Linear Model (GLM) implemented in SPM8 was used to derive task-fMRI activation maps. The task activation maps were confirmed by an expert neuroradiologist as consistent with the information provided during presurgical planning. The tumor boundaries for each patient were manually delineated by a medical fellow using the MIPAV software. Figure 2 shows language areas (red) for two separate subjects to illustrate the heterogeniety of our cohort.

Implementation Details and Evaluation Criteria: We parcellate our rs-fMRI data using the Craddocks atlas [14] with the cerebellar regions removed due to inconsistent acquisition ($N = 384$). Due to different patients performing different tasks, we train and test four separate GNNs, one for language identification and the rest for each motor task. We assign a parcel to the eloquent cortex if a majority of its voxels coincided with the ground truth task activations. We employ a ten-fold cross validation for the language experiment, and a five-fold cross validation for the motor experiments, as we observed the motor GNNs overfit more easily. For language, we stratify our folds by ensuring at most one bilateral language subject is in each fold. We report eloquent class accuracy as well as overall accuracy for each method that reflect a viable trade-off between true positive rate (TPR) and true negative rate (TNR). We compute and report area under the curve (AUC) by varying hyperparameter settings to approximate ROC. We consider eloquent vs. not eloquent for each ROC statistic reported. We maintain the same hyperparameter values across each of the four experiments. Tumor class accuracy is not reported, as both the MLP and GNN achieved near perfect ($\approx$.995) accuracy due to the assumptions of our setup.

Table 1. Node identification statistics for motor tasks.

Task	Method	Motor	Overall	Sensitivity	Specificity	AUC
Foot	Linear SVM	0.48	0.52	0.46	0.49	0.52
	RF	0.36	0.77	0.34	**0.86**	0.59
	MLP	0.73	0.76	0.63	0.75	0.74
	GNN	**0.84**	**0.81**	**0.78**	<u>0.79</u>	**0.81**
Tongue	Linear SVM	0.58	0.59	0.57	0.55	0.56
	RF	0.42	0.77	0.34	**0.92**	0.65
	MLP	0.75	0.78	0.68	0.77	0.75
	GNN	**0.87**	**0.84**	**0.82**	<u>0.80</u>	**0.83**
Finger	Linear SVM	0.60	0.62	0.56	0.58	0.57
	RF	0.49	0.80	0.43	**0.92**	0.69
	MLP	0.82	0.76	0.78	0.73	0.80
	GNN	**0.88**	**0.87**	**0.84**	<u>0.83</u>	**0.86**

3.1 Motor Class Identification

The motor identification results are reported in Table 1. As seen, our GNN overall outperforms all baselines in nearly all metrics. This notable performance is especially highlighted in the AUC column, as our method has the best trade-off between TPR and FPR. Our results suggest that approaching this problem with a deep learning framework is favorable, as both neural networks outperform the traditional machine learning baselines. Furthermore, we show a marked increase in performance employing the specialized convolutional filters. This suggests that our network learns a more discriminative representation of eloquent cortex at rest than the MLP. Due to different patient subcohorts performing different tasks, we train and test on each motor task separately. Figure 3 shows a coronal view of ground truth (red) and predicted (blue) motor regions for the foot (left) and tongue (right) tasks in one patient. As seen, our network is able to pinpoint both the midline and the peripheral areas of the motor strip. Our performance suggests that our method is able to localize specific parts of the motor homonculus, which is important in a preoperative setting.

Table 2. Node identification statistics for language class ($N = 60$).

Method	Language	Overall	Sensitivity	Specificity	AUC
Linear SVM	0.56	0.52	0.55	0.49	0.53
RF	0.37	0.77	0.33	**0.89**	0.63
MLP	0.66	0.73	0.61	0.76	0.70
GNN	**0.74**	**0.86**	**0.70**	<u>0.77</u>	**0.76**

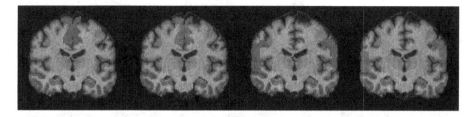

Fig. 3. Ground truth (red) and predicted (blue) motor regions for foot (left) and tongue (right) tasks in one patient. (Color figure online)

3.2   Language Class Identification

Table 2 reports the language identification performance across all methods. Once again, our GNN outperforms the baselines in nearly all methods, with the most notable gains in language accuracy and AUC. The specificity of our GNN is lower than expected due to the hemispheric symmetry of rs-fMRI data. We saw that the most frequent misclassification from our model was assigning contralateral parcels to the language class. We ran two experiments to probe whether our GNN is learning connectivity patterns associated with language rather than memorizing node locations. In Fig. 4 (left), we plot the histogram of true (pink) vs. predicted (blue) language parcels. The x-axis shows how frequently a certain group of parcels was assigned to language in the ground truth and predicted labels. Each bin represents a different group of parcels. Highlighted by the red box, we see that our model tends to overpredict the language class. We assert that this overprediction is viable due to the demands of the clinical application of our work. We compared the GNN output with seed based correlation analysis (SBA), where the "seed" for each patient is selected based on the ground truth task-fMRI activations. The average rs-fMRI time course within the seed location is correlated with each of the average time courses defined by our parcellation. The correlation maps are thresholded at $\rho > 0.6$ to retain only the strong associations. Figure 4 (right) shows a representative example, where the color of each parcel represents the strength of the connection; red is closer to 0.6 and yellow is closer to 1. Highlighted by the white arrows, there is right-hemisphere over prediction (blue) from our model. However, as shown by the SBA map, these right-hemisphere parcels have high resting-state connectivity with the seed average time course. Our GNN achieves a median of **0.81** dice overlap between the predicted language areas and the seed based correlation maps.

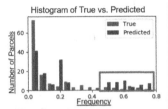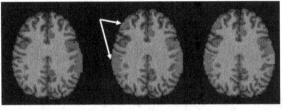

Fig. 4. Left: Histogram of true (pink) vs predicted (blue) language parcels for frequency > 0. Right: Arrows show overprediction overlaps with seed based maps. (Color figure online)

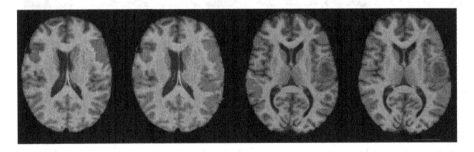

Fig. 5. Ground truth (red) and predicted (blue) for two separate subjects. All bilateral subjects were held out of training. (Color figure online)

Bilateral Language Identification: Our final experiment evaluates whether the GNN can recover a bilateral language network, even when this case is not present in the training data. Here, we trained the model on 55 left-hemisphere language network patients and tested on the remaining 5 bilateral subjects. Our model correctly predicted bilateral parcels in all five subjects. Figure 5 shows ground truth (red) and predicted language maps (blue) for two bilateral subjects. The median language class accuracy for these five cases was **0.62**. Empirically, this is slightly lower than reported in Table 2 due to the lack of training information. This experiment shows that our network learns connectivity patterns, rather than just spatial locations, of the language network.

4   Conclusion

We have demonstrated a GNN approach to identify the language and motor areas of eloquent cortex in brain tumor patients using rs-fMRI connectivity. Our model learns the resting-state functional signature of both the language and motor network within this tumor cohort by leveraging specialized convolutional filters that encode edge-node relationships within similarity matrices. With higher AUC for eloquent cortex detection, we prove that the features extracted from our GNN are more informative for this task than standard graph theoretic features and features extracted from a MLP. For language, we show that our model can correctly identify bilateral language networks even when trained on only unilateral network cases. Future work will decouple the lateralization problems in detecting language. We aim to add a separate network to our model that will determine which hemisphere(s) the language network is present in. We also aim to extend this work to simultaneously classify language and motor areas in one neural network, rather than training and testing these tasks separately.

Acknowledgements. This work was supported by the National Science Foundation CAREER award 1845430 (PI: Venkataraman) and the Research & Education Foundation Carestream Health RSNA Research Scholar Grant RSCH1420.

References

1. Duffau, H.: Lessons from brain mapping in surgery for low-grade glioma: insights into associations between tumour and brain plasticity. Lancet Neurol. **4**(8), 476–486 (2005)
2. Sair, H.I., et al.: Presurgical brain mapping of the language network in patients with brain tumors using resting-state fMRI: comparison with task fMRI. Hum. Brain Mapp. **37**(3), 913–923 (2016)
3. Langs, G., et al.: Functional geometry alignment and localization of brain areas. In: Advances in Neural Information Processing Systems, pp. 1225–1233 (2010)
4. Nandakumar, N., et al.: Defining patient specific functional parcellations in lesional cohorts via Markov random fields. In: Wu, G., Rekik, I., Schirmer, M.D., Chung, A.W., Munsell, B. (eds.) CNI 2018. LNCS, vol. 11083, pp. 88–98. Springer, Cham (2018). https://doi.org/10.1007/978-3-030-00755-3_10
5. Gohel, S., et al.: Resting-state functional connectivity of the middle frontal gyrus can predict language lateralization in patients with brain tumors. Am. J. Neuroradiol. (2019)
6. Hacker, C.D., et al.: Resting state network estimation in individual subjects. Neuroimage **82**, 616–633 (2013)
7. Lee, M.H., et al.: Clinical resting-state fMRI in the preoperative setting: are we ready for prime time? Top. Magn. Reson. Imaging: TMRI **25**(1), 11 (2016)
8. Kamnitsas, K., et al.: Efficient multi-scale 3D CNN with fully connected CRF for accurate brain lesion segmentation. MedIA **36**, 61–78 (2017)
9. Khosla, M., Jamison, K., Kuceyeski, A., Sabuncu, M.R.: 3D convolutional neural networks for classification of functional connectomes. In: Stoyanov, D., et al. (eds.) DLMIA/ML-CDS -2018. LNCS, vol. 11045, pp. 137–145. Springer, Cham (2018). https://doi.org/10.1007/978-3-030-00889-5_16

10. Kawahara, J., et al.: BrainNetCNN: convolutional neural networks for brain networks; towards predicting neurodevelopment. NeuroImage **146**, 1038–1049 (2017)
11. Opsahl, T., et al.: Node centrality in weighted networks: generalizing degree and shortest paths. Soc. Netw. **32**(3), 245–251 (2010)
12. Suresh, S., et al.: Risk-sensitive loss functions for sparse multi-category classification problems. Inf. Sci. **178**(12), 2621–2638 (2008)
13. Jack Jr., C.R., et al.: Sensory motor cortex: correlation of presurgical mapping with functional mr imaging and invasive cortical mapping. Radiology **190**(1), 85–92 (1994)
14. Craddock, R.C., et al.: A whole brain fmri atlas generated via spatially constrained spectral clustering. Hum. Brain Mapp. **33**(8), 1914–1928 (2012)

Graph Morphology-Based Genetic Algorithm for Classifying Late Dementia States

Oumaima Ben Khelifa[1,2] and Islem Rekik[2(✉)]

[1] BASIRA Lab, Faculty of Computer and Informatics,
Istanbul Technical University, Istanbul, Turkey
[2] National School of Engineers of Sousse, University of Sousse, Sousse, Tunisia
irekik@itu.edu.tr

Abstract. Early diagnosis of neurological diseases such as Alzheimer's disease (AD) is extremely vital for patient treatment. Analyzing the human brain connectivity is a popular approach in investigating the relationship between the brain morphology, structure, and function and the emergence of neurological diseases. However, extracting relevant diagnostic information from the connectome is still one of the most challenging problems. Many works have thoroughly studied the connectional map of the brain, however, to the best of our knowledge, no previous study had used *graph morphology* to rigorously explore the topological properties of the human connectome. In this paper, we propose a novel graph morphology-based genetic algorithm (GMGA) to mine the brain network and extract the most relevant connections for disordered brain state stratification. *First*, we define our graph morphological structural operators (SE) and design a subgraph matching technique for matching a particular graph-based SE with an input brain connectome. *Second*, we propose GMGA which identifies the optimal sequence of morphological operations using a predefined structural element for distinguishing between two brain states (e.g., late mild cognitive impairment (LMCI) vs Alzheimer's disease (AD)). *Last*, we train a linear classifier in a K-fold cross-validation fashion using the *morphed* brain graphs given the *optimal* learned morphological operator sequence. Our experimental results demonstrate a significant gain in classification performance between LMCI and AD groups in comparison with baseline methods. This work constitutes the first proof-of-concept of the merit of graph morphology in decoding the healthy and disorder brain connectomes.

Keywords: Brain connectivity · Brain dementia state classification · Graph matching · Genetic algorithm · Morphological brain networks

1 Introduction

The human brain network connectivity has long been the subject of numerous studies [1,2,13]. In fact, understanding and mapping the human connectome is

© Springer Nature Switzerland AG 2019
M. D. Schirmer et al. (Eds.): CNI 2019, LNCS 11848, pp. 21–31, 2019.
https://doi.org/10.1007/978-3-030-32391-2_3

of great importance in detecting the early signs of various clinical disorders of the brain (e.g., Alzheimer's Disease (AD) [14] and late mild cognitive impairment (LMCI)). In modern neuroscience, graphs are used to model the interactions between the different regions of the brain considering their abilities in capturing the topology and dynamics of the brain networks in both healthy and disordered states. That is, the nodes of the graphs usually represent the brain regions while the edges represent either anatomical, functional, morphological or effective connections. Understanding the topological complexity of the human connectome can be particularly relevant in detecting the atypical changes in the brain connectivity. These changes are critical in the diagnosis of neurodegenerative diseases.

To identify the diagnostic biomarkers for these diseases, several works relied on vectorizing the brain graph in Euclidean space in order to enable access to the full repository of machine-learning methods. To this end, graph embedding and other kernel methods are commonly used in connectomics. In [11], two vector-space embeddings are used to decode the brain connectivity: direct connection label sequence embedding and dissimilarity-based embedding. [8] uses a multivariate pattern analysis (MVPA) method to distinguish patients with social anxiety disorder (SAD) from healthy controls. While in [16] a novel network construction method based on a graph regularized weighted sparse model is proposed for the study of brain functional networks. However, despite their empirical effectiveness, these methods fail to fully exploit the topological properties of the brain and might cause losing valuable information about the brain as a connectional highly-interative and complex construct. Thus, to investigate the brain graph topology, we propose a novel method rooted in the *theory of graph morphology*, which remains largely unexplored in the field of connectomics. By emphasizing the topological properties of the brain network, we will be able to eventually model how a neurological disorder can change the brain connectional map.

Graph morphology was first introduced in [5,15] and similar morphological operators were used in image processing [3]. However, to the best of our knowledge, no previous graph morphology methods were used on the brain graph. To this end, we propose a novel graph morphology-based method that scour the human connectome in order to accentuate the most relevant connections. *First*, we design a set of morphological structural elements (SE) composed of a center node and a set of subordinate nodes, then propose a subgraph matching technique to match a given SE to a whole brain graph of interest. Each structural element acts as a probe to extract diagnostic information from the brain network. Since the brain connectome is a *weighted* graph, our proposed subgraph matching technique takes into account the connectional strength (i.e., the weights of the edges of the graph). *Second*, we define the set of morphological operators which will be applied to the brain graph using a SE of interest: (i) the vertex-edge dilation and (ii) the vertex-edge erosion. *Last*, we present a graph morphology-based genetic algorithm (GMGA) that searchs for the fittest sequence of SE-based morphological operations that can mimic the alterations of the brain connectivity caused by the neurodegenerative diseases. Particularly, by morphing each brain graph using the learned optimal morphological sequence, we train a linear classifier in a K-fold cross-validation manner to classify patients diagnosed with LMCI and AD.

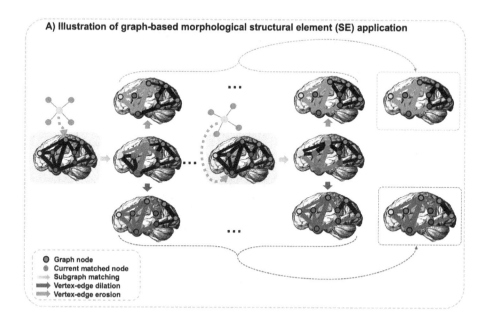

Fig. 1. *Example of morphological operations on a brain network: vertex-edge dilation (red) and vertex-edge erosion (blue).* The structural element (SE) here is composed of a center node (yellow node) connected to four subordinate nodes (green nodes). The first step is matching the SE to the brain graph where we first activate a matched node to the center node (the pink node for the first subgraph and the red node for the last subgraph). Then we match the other four nodes to the nodes in the brain graph with the highest brain connectivities weights (i.e., thickest edges). Once we match the SE to the brain graph, we perform a vertex-edge erosion (blue) and a vertex-edge dilation (red). Finally, after dilating (resp., eroding) all the matched subgraphs, we obtain the result of the SE-based dilation (resp., erosion) of the whole graph by merging together the morphed subgraphs. (Color figure online)

2   Method

Mathematical morphology is a well-defined theory [6], where structural elements act as probes to extract structural information from geometric spaces, nesting images, graphs and other data structures. In this paper, we focus on graph morphology and propose a novel approach for applying different morphological operators to brain networks in order to better capture its topological complexity. In this section, we introduce the different steps of our method, beginning with

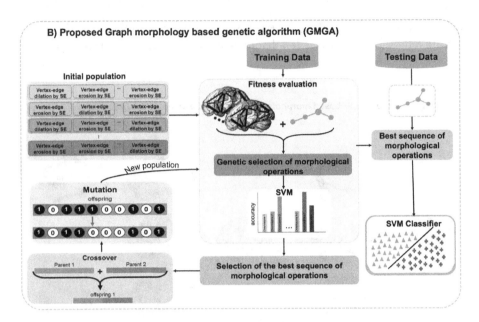

Fig. 2. *Overview of the proposed pipeline for the graph morphology based genetic algorithm (GMGA).* We first initialize a population of chromosomes, where each chromosome stands for a sequence of morphological operations that will act on the training brain graphs using a predefined SE. Each chromosome will then be graded according to their accuracy score. Next, the fittest chromosomes are selected then added to the mating pool. A crossover will then occur resulting in the production of the offspring. One of the offspring will be randomly selected to undergo a mutation. Lastly a new population will replace the previous one consisting of the new offspring and their parents. This cycle will repeat until a termination criterion is reached. This will result in an optimal sequence of morphological operations that will then be applied to the training brain graphs which in turn will be used to train an support vector machine (SVM) classifier. Last, we similarly morph the testing brain graph and feed it to SVM for label prediction.

the first step, subgraph matching, since each structural element is defined as a subgraph. Then we define our morphological operators the vertex-edge erosion and the vertex-edge dilation, acting on the brain graph using a SE of interest. And finally, we detail our proposed graph morphology-based genetic algorithm (GMGA) for brain state classification.

2.1   Proposed Graph-Based Structural Element Matching Technique

Similar to classical morphology, the first step is defining the structural element (SE). In our case the structural element is a structural subgraph. Each structural subgraph has a simple and small structure: a center node which anchors the SE matching process to the brain graph (the yellow node in the green SE in Fig. 1) and a set of subordinate nodes (green neighboring nodes surrounding the center node in Fig. 1). Let G be an undirected graph $G = (G^V, G^E)$, where G^V denotes the set of vertices and G^E the set of edges in G. Let G' denote a subgraph of G if $G'^V \in G^V$ and $G'^E \in G^E$.

However, since we are working on brain graphs, our subgraph matching technique depends on the edge weights (i.e., connectivity strength between different anatomical regions of interest (ROIs) in the input brain network). Hence, we propose a novel graph-based SE matching strategy, which matches an unweighted subgraph to a weighted graph. First, given a SE, we loop over all the nodes of the graph and activate the nodes which align with the SE topological structure centered at the SE center (Fig. 1) –i.e., their local structure patterns are identical. Once the nodes are activated, we then sort the subordinate matched nodes in the weighted brain graph according to their connectivities weights. Next, we activate the brain graph edges with the strongest connections matching the input SE subgraph. Figure 1–A illustrates these steps where first the pink node is activated which means that the SE structure centered at the yellow node aligns with the pink node. In other terms, the SE structure was successfully identified at the pink node. Once activated, a sorting of the edges occurs based on their connectivity weights. Next, given the matched list of the sorted edges to the SE of interest, the nodes with the strongest connections with the pink node are matched. Notice that even though the light blue node is a possible candidate we do not consider it as a match since the other four nodes have stronger connections with the pink node. Note that this is different from existing graph morphology methods [3,5] that do not take into account the weights of the graph.

Throughout the remainder of this paper, we shall denote by $M(G|SE)$ the set of subgraphs resulting from matching the structural element SE to the graph G.

2.2 Morphological Operators

After matching our structural element to the brain graph of interest, we define a set of morphological operators, which will act on the resulting subgraphs in $M(G|SE)$ (Fig. 1). In this study, we mainly focus on the connectivity of the network. Therefore, it is more convenient to consider morphological operators that affect the edges of the graph (not the nodes).

- **Vertex-edge dilation:**

We denote by δ the vertex-edge dilation that goes from G^V to G^E, mapping the set of vertices to the set of edges that are connected to at least one of these vertices. Let $A = (A^V, A^E)$ denote a subgraph of G. The vertex-edge dilation of A is defined as:

$$\delta(A) = \{e_{u,v} \in G^E \mid \text{either u} \in A^V \text{ or v} \in A^V\}.$$

Respectively, the vertex-edge dilation of the graph G by the structural element SE is the union of all the vertex-edge dilations of the subgraphs in $M(G|SE)$ (Fig. 1):

$$\delta(G|SE) = \{\delta(A) \mid A \in M(G|SE)\}$$

- **Vertex-edge erosion:**

We denote by ϵ the vertex-edge erosion that goes from G^V to G^E, mapping the set of vertices to the set of edges whose two extremities are included in the same set of vertices. Let $A = (A^V, A^E)$ be a subgraph of G. The vertex-edge erosion of A is defined as:

$$\epsilon(A) = \{e_{u,v} \in G^E \mid u \in A^V \text{ and v } \in A^V\}.$$

Similar to the vertex-edge dilation of G by SE, the vertex-edge erosion of the graph G by the structural element SE is the union of all the vertex-edge erosions of the subgraphs in $M(G|SE)$ (Fig. 1):

$$\epsilon(G|SE) = \{\epsilon(A) \mid A \in M(G|SE)\}$$

We note that both the vertex-edge erosion and the vertex-edge dilation are applied iteratively. Therefore, the order in which we apply the morphological operations on the matched brain subgraphs in $M(G|SE)$ will not affect the morphing of the graph G. This can be explained by the fact that we *independently* apply the SE-based morphological operation to each matched subgraph in the input brain graph, then merge all morphed graphs. After applying the morphological operations to the graph, in our case on the brain graph, the output graph is then transformed and only the edges accentuated by these operations remain.

2.3   Graph Morphology Based Genetic Algorithm (GMGA)

Inspired by the evolution theory in biology, genetic algorithms (GAs) [7] were introduced as a population-based optimization process. A typical GA maintains a population of N individuals (chromosomes) by relying on biological operators such as mutation, crossover and selection. In our study, each chromosome stands for a sequence of SE-based morphological operations. Specifically, we seek to find the best one that can mimic how a disorder can alter the connectivity of the brain network. We illustrate in Fig. 2 the key components of the proposed pipeline, which we detail below.

Initialization. In this step, an initial population of N chromosomes is set as an input where each chromosome is represented by a binary code. We also have to specify the structural element used and the number of iterations which will work as the termination criterion.

Fitness Evaluation. For each chromosome in the population, a fitness score is associated. This is the most critical step. Here, each chromosome stands for a sequence of SE-based morphological operations, which is performed on our brain graphs. Then we use a support vector machine (SVM) linear classifier to rank each chromosome according to its accuracy scores.

Selection. Based on their fitness scores, we select the best chromosomes and add them to the mating pool allowing them to pass their genes to the next generation.

Crossover. Immediately after selecting the fittest chromosomes, every two "parents" are paired together to produce an offspring. Each offspring inherits half of his genes from each parent.

Mutation of the Offspring. Once the offspring are produced, one will be randomly selected to undergo a mutation. One gene is then randomly selected to change their status. This step will ensure the diversity in the population to avoid the premature convergence.

Once this step is complete, the new population, consisting of parents and offspring, replaces the previous population and our algorithm resumes the fitness evaluation. It is important that the new generation contains the best chromosomes of the previous generation (the parents). This will ensure that the performance of this population will not get worse since there is no guarantee that the new chromosomes (i.e., defined SE-based morphological operator sequence) will give a better fitness score. After a predefined number of iterations, the optimal sequence of morphological operations disentangling two brain groups is generated. This sequence is encoded in the chromosome with the highest fitness score and highest ability to classify both training groups. We also note that the optimal SE-based morphological operator sequence is learned in K-fold cross-validation fashion. Once the optimal sequence is identified using the training brain graph samples, we morph all training graphs to train a linear classifier. In the testing stage, we apply the same sequence to morph the testing brain graphs, then input them to the trained classifier for predicting the final brain state of each testing subject (i.e., label).

3   Results and Discussion

Dataset. We used 11-fold cross validation to evaluate our proposed method on 77 subjects (41 AD and 36 LMCI) from ADNI GO public dataset each with structural T1-w MR image. Data used in the preparation of this article were obtained from the Alzheimer's Disease Neuroimaging Initiative (ADNI) database (adni.loni.usc.edu). The ADNI was launched in 2003 as a public-private partnership, led by Principal Investigator Michael W. Weiner, MD. The primary goal of ADNI has been to test whether serial magnetic resonance imaging (MRI), positron emission tomography (PET), other biological markers, and clinical and neuropsychological assessment can be combined to measure the progression of mild cognitive impairment (MCI) and early Alzheimer's disease (AD). We use FreeSurfer processing pipeline to reconstruct both right (RH) and lef (LH) cortical hemispheres for each subject from T1-w MRI. Then each cortical hemisphere was parcellated into 35 cortical ROIs using Desikan Killiany cortical atlas. Next,

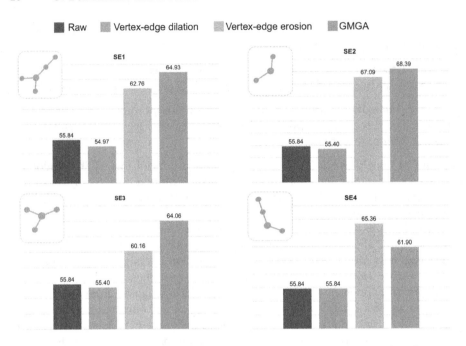

Fig. 3. Evaluating the performance of our proposed graph morphology-based genetic algorithm (GMGA) for classifying LMCI and AD patients using four different structural elements on morphological brain graphs measured using maximum principal curvature.

for each subject, we generate a morphological brain network using the maximum principal curvature as a metric to quantify the dissimilarity in morphology between pairs of ROIs [4,9,10,12].

Parameters. We note that we applied thresholding to this dataset before the application of GMGA to create sparser brain graphs by eliminating the weaker connections (i.e., where the weights of the edges are below a certain threshold). Specifically, we evaluated GMGA on three datasets, each sparsified at a specific threshold: (1) the mean μ of all the weights of the brain graph, (2) $\mu + \eta$, where η denotes the standard deviation of the data brain edge weights, and (3) $\mu - \eta$. Next, we report in Figs. 3 and 4 the averaged classification results across the three thresholded connectomic datasets. For GMGA, each population P contains $N = 6$ chromosomes and each chromosome has 6 genes (i.e., morphological operators –not necessarily different) with the termination criterion being 15 iterations. As shown in Fig. 3, our method boosts the classification results in comparison with the raw data.

Varying Structural Elements. We evaluated our method using different structural elements as shown in Fig. 3. It can easily be seen that the performance of this method depends mainly on the choice of the SE.

Comparing the performance of each SE, we report that SE2 had the highest accuracy score 68.39 while SE4 had the lowest 61.9. However, we should state that all four of these SEs improved the accuracy by a large margin. It is also interesting to note that the vertex-edge erosion can yield good result on its own; however, a much better result can be obtained by sequencing many vertex-edge erosions and vertex-edge dilations. Furthermore, the fact that very good results are obtained by applying a sequence of morphological operations shows that our technique is able to accentuate the critical connections in the brain network which are affected by the neurodegenerative diseases.

Varying the Number of Genes. We also evaluated how different number of genes (i.e., the number of morphological operations) can affect the accuracy as shown in Fig. 4. The population with chromosomes consisting of 6 genes yields better classification accuracy than those with only 4 genes for both structural elements SE1 and SE2. However, when we increase the number of genes to 8, we find that the accuracy of our algorithm when applied with SE1 increases while the accuracy of our algorithm when applied with SE2 decreases. Moreover, the longer our chromosomes are the longer time needed for our algorithm. Therefore we conclude that a population with chromosomes consisting of 6 genes has better empirical performance. However, even though our algorithm has proven to predict the best sequence of morphological operations given a predefined structural element, the main limitation of our proposed work is that we predefined the structural elements. In our future work, we will learn to identify the most relevant SE for the target classification task. This is particularly important since the performance of our algorithm mainly depends on the choice of the structural

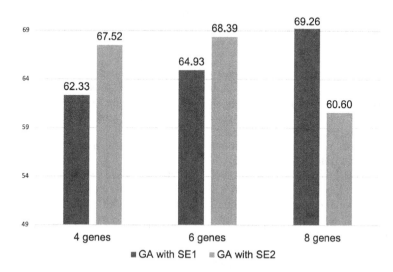

Fig. 4. Comparing between the performance of our proposed algorithm using the previous first two structural elements SE1 and SE2 with different numbers of genes (number of morphological operations).

element. Furthermore, there are several future directions to build on this seminal work including mixing structural elements to identify the optimal morphological sequence as well as defining weighted structural elements.

4 Conclusion

In this work, we proposed a novel graph morphology-based genetic algorithm (GMGA) for brain state classification using morphological connectomic data. We conducted experiments on the ADNI GO public dataset and reported boosted classification results when integrating graph morphology theory with genetic algorithm for learning how to optimally morph disordered and healthy brain graphs to accentuate their connectional differences. By learning the optimal sequence of morphological operations for different structural elements, we were able to investigate brain topological changes at different sub-graphic scales. Our work shows that graph morphology can be a valuable approach for investigating the topology of brain networks. There are several future directions we intend to explore including the joint learning of the optimal structural elements along with the optimal set of morphological operators.

References

1. Bullmore, E., Bassett, D.: Brain graphs: graphical models of the human brain connectome. Annu. Rev. Clin. Psychol. **7**, 113–140 (2011)
2. Bullmore, E., Sporns, O.: Complex brain networks: graph theoretical analysis of structural and functional systems. Nat. Neurosci. **10**, 186–198 (2009)
3. Cousty, J., Najman, L., Serra, J.: Some morphological operators in graph spaces. In: Wilkinson, M.H.F., Roerdink, J.B.T.M. (eds.) ISMM 2009. LNCS, vol. 5720, pp. 149–160. Springer, Heidelberg (2009). https://doi.org/10.1007/978-3-642-03613-2_14
4. Dhifallah, S., Rekik, I., Alzheimer's Disease Neuroimaging Initiative, et al.: Clustering-based multi-view network fusion for estimating brain network atlases of healthy and disordered populations. J. Neurosci. Methods **311**, 426–435 (2019)
5. Heijmans, H., Nacken, P., Toet, A., Vincent, L.: Graph morphology. J. Vis. Commun. Image Represent. 24–38 (1990)
6. Heijmans, H., Vincenty, L.: Graph morphology in image analysis. Math. Morphol. Image Process. **34**, 171–203 (1992)
7. Holland, J.H.: Genetic algorithms. Sci. Am. **267**(1), 66–72 (1992)
8. Liu, F., et al.: Multivariate classification of social anxiety disorder using whole brain functional connectivity. Brain Struct. Funct. **220**, 101–115 (2015)
9. Mahjoub, I., Mahjoub, M.A., Rekik, I.: Brain multiplexes reveal morphological connectional biomarkers fingerprinting late brain dementia states. Sci. Rep. **8**(1), 4103 (2018)
10. Nebli, A., Rekik, I.: Gender differences in cortical morphological networks. Brain Imaging Behav. 1–9 (2019)
11. Richiardi, J., Ville, D.V.D., Riesen, K., Bunke, H.: Vector space embedding of undirected graphs with fixed-cardinality vertex sequences for classification. In: 20th International Conference on Pattern Recognition, pp. 902–905 (2010)

12. Soussia, M., Rekik, I.: Unsupervised manifold learning using high-order morphological brain networks derived from T1-w MRI for autism diagnosis. Front. Neuroinform. **12** (2018)
13. Sporns, O.: The human connectome: a complex network. Ann. N.Y. Acad. Sci. **1224**, 109–125 (2011)
14. Tijms, B., et al.: Alzheimer's disease: connecting findings from graph theoretical studies of brain networks. Neurobiol. Aging **34**, 2023–2036 (2013)
15. Vincent, L.: Graphs and mathematical morphology. Signal Process. **16**, 365–388 (1989)
16. Yu, R., Qiao, L., Chen, M., Lee, S.W., Fei, X., Shen, D.: Weighted graph regularized sparse brain network construction for MCI identification. Pattern Recogn. **90**, 220–231 (2019)

Covariance Shrinkage for Dynamic Functional Connectivity

Nicolas Honnorat[1]([✉]), Ehsan Adeli[2], Qingyu Zhao[2], Adolf Pfefferbaum[1,2], Edith V. Sullivan[2], and Kilian Pohl[1,2]

[1] SRI International, Menlo Park, CA, USA
nicolas.honnorat@sri.com
[2] Department of Psychiatry and Behavioral Sciences, Stanford University, Palo Alto, CA, USA

Abstract. The tracking of dynamic functional connectivity (dFC) states in resting-state fMRI scans aims to reveal how the brain sequentially processes stimuli and thoughts. Despite the recent advances in statistical methods, estimating the high dimensional dFC states from a small number of available time points remains a challenge. This paper shows that the challenge is reduced by *linear covariance shrinkage*, a statistical method used for the estimation of large covariance matrices from small number of samples. We present a computationally efficient formulation of our approach that scales dFC analysis up to full resolution resting-state fMRI scans. Experiments on synthetic data demonstrate that our approach produces dFC estimates that are closer to the ground-truth than state-of-the-art estimation approaches. When comparing methods on the rs-fMRI scans of 162 subjects, we found that our approach is better at extracting functional networks and capturing differences in rs-fMRI acquisition and diagnosis.

1 Introduction

The development of resting-state functional MRI (rs-fMRI) has provided a way to measure spontaneous brain activity across the brain, in vivo and in real-time [1]. Similar patterns of spontaneous brain activity across brain regions are referred to as functional connectivity. Traditionally, these patterns were assumed to be static over the acquisition time of the rs-fMRI, but recent work suggests otherwise [2]. Called Dynamic functional connectivity (dFC) patterns, they are often clustered across time frames of the rs-fMRI series to identify recurring connectivity patterns or 'mind states' [7]. Analysis of these mind states has given new insights into mental disorders and neurodevelopment [9].

According to a recent review [9], the most accurate method for estimating dFC is called Dynamic Conditional Correlation (DCC) [6]. DCC filters the rs-fMRI signal via the generalized autoregressive conditional heteroskedasticity approach (GARCH) and defines a dFC for each time point of the blood-oxygen-level dependent (BOLD) time series as the remaining global connectivity at that

© Springer Nature Switzerland AG 2019
M. D. Schirmer et al. (Eds.): CNI 2019, LNCS 11848, pp. 32–41, 2019.
https://doi.org/10.1007/978-3-030-32391-2_4

time. This global connectivity is captured by a covariance matrix across all voxels or regions of interest estimated using the Exponentially Weighted Moving Average method (EWMA) [6]. EMWA, however, was not designed for estimating large covariance matrices from a small number of samples (or time points) as it is the case in this application. In addition, the estimation is computationally expensive preventing this type of analysis to be directly applied to the entire high resolution fMRI of large data sets.

In this work, we tackle these issues by introducing an accurate and efficient implementation of EWMA. Specifically, we first show that EWMA computes weighted covariance matrices, where the weights are defined according to a continuous sliding window. We then improve the accuracy of that weighted covariance estimation via linear *covariance shrinkage* [5], a statistical method designed for the estimation of large covariance matrices in low sample size settings. Finally, we reformulate the approach to cluster dFCs without having to explicitly compute the covariances themselves. Thus, we can efficiently identify mind states on large data sets. Compared to DCC, the dFC estimated by our approach is closer to the ground-truth on a synthetic data set. When applied to a rs-fMRI data set of 162 subjects, our approach is better at estimating functional networks and at capturing differences in MRI acquisition, and between healthy controls and those with alcohol use disorder.

2   Covariance Shrinkage for dFC

2.1   EWMA-Based dFC Estimation Using Continuous Sliding Windows

Let n be the number of time points and p the number of voxels or regions of interests of a rs-fMRI, then we denote with $X = [x_1, ..., x_n]$ the $p \times n$ matrix storing all the BOLD measurements. We assume that these measurements have been processed by a neuroimaging pipeline including standard motion correction, temporal band-pass filtering, and BOLD signal normalization. The entries of each row of X are therefore assumed to have a zero mean and unit variance. To relate EWMA to a continuous sliding window approach, we introduce a continuous sliding window $w_t := [w_t(1), ..., w_t(n)]$ for each time point t. The entries of w_t are positive and add up to 1.

$$
\begin{aligned}
w_1 &:= (1, 0, 0, .., 0),\\
w_t(i) &:= \begin{cases} 1 - \theta & \text{if } i = t \\ \theta\, w_{t-1}(i) & \text{otherwise} \end{cases},
\end{aligned}
\tag{1}
$$

where $\theta \in [0, 1]$ specifies the weight of previous time points. We then compute the weighted covariance matrix with respect to time point t as

$$
\begin{aligned}
C_t &:= \sum_{i=1}^{n} w_t(i) \left(x_i - \sum_{j=1}^{n} w_t(j)x_j \right) \left(x_i - \sum_{j=1}^{n} w_t(j)x_j \right)^T\\
&= \sum_{i=1}^{n} w_t(i)x_i x_i^T - \sum_{ij} w_t(i)w_t(j)x_i x_j^T.
\end{aligned}
\tag{2}
$$

This covariance computation is mathematically equivalent to the recursive computation carried out by the EWMA in [6].

2.2   Linear Covariance Shrinkage

This new mathematical formulation offers a way to improve the EWMA estimations used for fMRI connectivity analysis. As mentioned, computing $C_t \in \mathbb{R}^{p \times p}$ is not reliable as the number of measurements p is much larger than the number of samples or time points n. Linear covariance shrinkage [5] mitigates this issue by replacing the empirical covariance with a linear combination between itself and its trace $Tr(\cdot)$:

$$C_t^* = (1 - \lambda_t)C_t + \lambda_t \frac{Tr\,(C_t)}{p} I. \tag{3}$$

Under the assumption that the measurements are Gaussian distributed and $w_t(i) = 1/n$, the optimal setting of the parameter λ can be directly computed from C_t and n according to the Oracle Approximating Shrinkage [3]:

$$\lambda_t = min\left(1, \frac{\left(1 - \frac{2}{p}\right)Tr\left(C_t^2\right) + Tr^2\left(C_t\right)}{\left(n + 1 - \frac{2}{p}\right)\left[Tr\left(C_t^2\right) - \frac{Tr^2(C_t)}{p}\right]}\right). \tag{4}$$

We found that the proof in [3] can be generalized to embed covariance shrinkage into EMWA (i.e., Eq. (2)) and thus drop the assumption that $w_t(i) = 1/n$. We replaced the original Wishart distribution moments in [3] with moments obtained for Wishart matrices weighted by w_t and we followed the derivations step by step. We finally found that Eq. (4) still holds, if one replaces the number of time points n with the *effective number of samples*:

$$n_w = \frac{\left(\sum_i w_t(i)\right)^2}{\sum_i \left(w_t(i)\right)^2}. \tag{5}$$

2.3   Efficient Implementation

Typical dFC analysis involves clustering the covariance matrices, which requires computing λ_t for each time point and ℓ_2 distance between each pair of covariances C_s^* and C_t^*. We found that the following computational trick can be used to efficiently compute λ and the ℓ_2 distance without computing the matrices themselves. Specifically, we introduce a matrix $K := X^T X \in \mathbb{R}^{n \times n}$, which is a very small matrix that can be computed at once, at the very beginning of the fMRI analysis. Denoting the entry-wise product between matrices with $\odot$, we found that Eq. (4) can be efficiently estimated with

$$\begin{aligned}
Tr\,(C_t) &= w_t^T Diag(K) - w_t^T K w_t, \\
Tr\,(C_t^2) &= w_t^T (K \odot K) w_t - 2w_t^T K Diag(w_t) K w_t + \left(w_t^T K w_t\right)^2, \\
Tr\,(C_s C_t) &= w_s^T (K \odot K) w_t - L_{st} + \left(w_s^T K w_t\right)^2, \\
\text{where } L_{st} &= w_s^T K Diag(w_t) K w_s + w_t^T K Diag(w_s) K w_t.
\end{aligned} \tag{6}$$

Now let $\gamma := \lambda_s Tr(C_s)/p$ and $\delta := \lambda_t Tr(C_t)/p$, the ℓ_2 distance between two covariances C_s^* and C_t^* is efficiently obtained by

$$
\begin{aligned}
||C_s^* - C_t^*||_2^2 &= ||(1-\lambda)C_s + \gamma I - (1-\mu)C_t - \delta I||_2^2 \\
&= 2(\gamma - \delta)(1-\lambda)Tr(C_s) + 2(\delta - \gamma)(1-\mu)Tr(C_t) \\
&\quad + (1-\mu)^2 Tr(C_t^2) - 2(1-\lambda)(1-\mu)Tr(C_s C_t) \\
&\quad + (1-\lambda)^2 Tr(C_s^2) + (\gamma - \delta)^2 p.
\end{aligned}
\tag{7}
$$

This trick, inspired from the Support Vector Machine literature [4], reduces the computational burden by several order magnitudes during our experiments. It provides us with a mean to process full-resolution scans in reasonable time, and spares a significant amount of computer memory by preventing the computation of the covariances matrices, which would have contained billions of entries.

We further improve the efficiency of the computations by exploiting the relation between consecutive EWMA time windows. More specifically, we recursively compute the traces of the covariance matrices (Eq. (6)) by introducing intermediate variables:

$$
\begin{aligned}
\alpha_t &:= w_t^T Diag(K), \\
\epsilon_t &:= K w_t, \\
\beta_t &:= w_t^T \epsilon_t,
\end{aligned}
\tag{8}
$$

so that $Tr\,(C_t) = \alpha_t - \beta_t$. Now let $K_{\cdot t}$ denote the column of K corresponding to time point t, then the algorithm for recursively computing the intermediate variables is as follows

$$
\begin{array}{ll}
\textbf{Initialization} & \textbf{Recursion} \\
\alpha_1 = K_{11} & \alpha_t = \theta\alpha_{t-1} + (1-\theta)\,K_{tt} \\
\epsilon_1 = K_{\cdot 1} & \epsilon_t = \theta\epsilon_{t-1} + (1-\theta)K_{\cdot t} \\
\beta_1 = K_{11} & \beta_t = \theta^2\beta_{t-1} + 2\theta\,(1-\theta)\,\epsilon_{t-1}(t) + (1-\theta)^2\,K_{tt}
\end{array}
\tag{9}
$$

Similarly, $Tr\,(C_t^2)$ and $Tr\,(C_s C_t)$ in Eq. (6) can also be efficiently computed based on the above intermediate variables. This strategy allows estimating dFCs from full resolution fMRI in a few minutes using a standard office computer instead of repeatedly computing C_t matrices which would require terabytes of memory. Furthermore, it allows clustering the dFC in reproducible transient mind states efficiently, by computing the distance matrix D across all dFCs

$$
D_{st} = ||C_s^* - C_t^*||_2^2,
\tag{10}
$$

and applying a clustering method to D. We establish the validity of this approach by computing such matrices D during our experiments, and extracting from these matrices a measurement indicative of the "clusterability" of the time points, which are potential biomarkers.

3 Experiments

3.1 Data Sets

The first data set consists of rs-fMRI of 162 subjects [8]: 18 were diagnosed with alcohol dependence (ALC) while the remaining samples were labeled as controls.

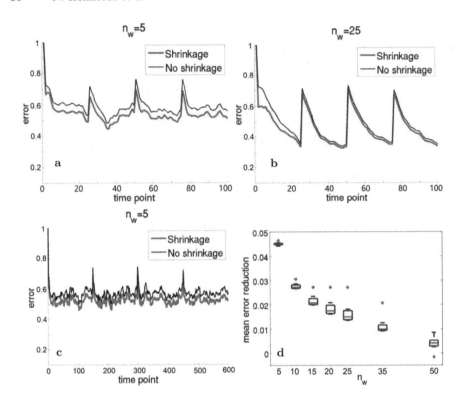

Fig. 1. Average of the Pearson distances measured for the data sets with 25 time points per brain state between the ground-truth and the dFCs estimated by EWMA, which we refer to as the "estimation error" for (a) $n_w = 5$, (b) $n_w = 25$. (c) for 150 time points per brain state and $n_w = 5$. (d) estimation error measured without covariance shrinkage minus the estimation error with covariance shrinkage. Covariance shrinkage consistently reduces the estimation error. This improvement is larger for small n_w, i.e., short time windows.

The total group comprising 73 female and 89 male participants, ranged in age from 23–80 years. Furthermore three different acquisition sequences were used to acquire the data. The fMRI repetition time (TR) of these protocols was 2 s for 61 subjects, 2.2 s for 67, and 2.648 s for the remaining 34. The processing of the rs-fMRI scan consisted of spatial smoothing with 5 mm FWHM, temporal de-trending, and band-pass filtering between 0.01 and 0.1 Hz. The processed scans were non-rigidly registered to an atlas of 111 regions of interest [10]. We computed the average BOLD signal inside each region to produce low-resolution time series. Furthermore, we kept the high-resolution time series, which consisted of the 175473 BOLD series within the gray matter. As a result, the spatial resolution of our high resolution series was thousand times finer than that of typical dFC studies [9,11].

We created a synthetic data set by first defining four distinct brain states. To do so, we computed for each subject the correlation matrix of the low-resolution time series. We selected for the ground-truth brain states the four correlation matrices with the largest ℓ_2 distances to each others. To create a synthetic rs-fMRI scan, we generated for each state a random band-passed signal by (i) creating a random timeseries, with a sampling rate of 0.5 Hz, by randomly sampling from the normal Gaussian distribution $N(0,1)$, (ii) band-pass filtering these time series using Butterworth filter of order 2 and a band-pass filter of $[0.01, 0.1]$ Hz, and (iii) multiplying the square root of the state correlation matrix with the filtered time series. Lastly, we concatenated the four random band-passed signals to create a synthetic rs-fMRI sequence. This process was repeated 10 times for each data set. We created 8 of these data sets with different number of time points per brain state: 25, 50, 75, 100, 125, 150, 200 and 250.

3.2   Implementations

The baseline for our experiments was defined by an implementation of DCC, which was applied to all our data sets. Specifically, we first filtered rs-fMRI time series via GARCH. Next, we estimated the dFC by applying our efficient implementation of EWMA without covariance shrinkage. Note, we ensured on the synthetic data set that the outcome of that implementation was exactly the same as the EWMA implementation originally proposed in [6]. This implementation was applied to each data set seven times with the effective number of samples n_w being $\{5, 10, 15, 20, 25, 35, 50\}$. We then repeated these computations a second time using our proposed approach, the efficient EWMA implementation with covariance shrinkage.

3.3   Findings on the Synthetic Data

Figure 1 plots the error of each approach with respect to the synthetic data set consisting of brain states of 25 time points and the effective sample size being (a) $n_w = 5$ and (b) $n_w = 25$. The error is defined by the Pearson distance between the estimated and ground-truth covariance matrices averaged across the corresponding 10 scans. Our proposed approach exhibits a lower error for almost all time points and the difference to the baseline was especially large for the smaller effective sample size. The results obtained with different number of time points per brain states were very similar, as shown for instance by Fig. 1(c). The box plot of Fig. 1(d) confirms this finding as we observe better estimations using covariance shrinkage for all the parameter values. The improvement is larger for the smallest n_w, which corresponds to the shortest dFC patterns. These results strongly advocate for the use of covariance shrinkage when using EWMA to estimate dFC.

3.4   Results on the Real rs-fMRI Data

Figure 2 plots the distance matrix D for both approaches with respect to the high resolution data and three effective number of samples n_w. For all three settings,

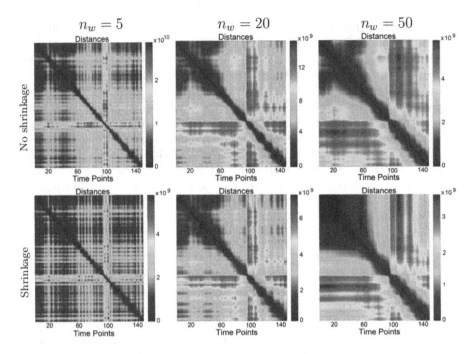

Fig. 2. The distance matrix D based on the dFC estimation of both approaches applied to the high resolution scans with different effective sample sizes n_w. Covariance shrinkage generates sharper distance matrices which would yield more reliable clusterings.

the pattern for the distance matrices computed by the proposed approach is sharper, especially for the smallest n_w. This observation suggests that clustering dFC into distinct mind states is easier when using covariance shrinkage.

We confirmed these qualitative findings by computing the quartile coefficient of dispersion (QCD) of all the distance matrices D obtained during our experiments. To compute the QCD, we first computed the first quantile $Q1$ and the third quartile $Q3$ of the distribution defined by the entries of D. We then defined

$$QCD := \frac{Q3 - Q1}{Q3 + Q1}. \tag{11}$$

We chose this metric as it is robust against outliers and is confined to $[-1, 1]$, which allows us to compare distance matrices of different magnitudes. More importantly, the larger the QCD value is, the easier it is to partition the matrix into distinct clusters, and the easier it is to separate dFCs into mind states. Figure 3(a) plots the QCDs of the two approaches when applied to the high resolution scans with the effective sample size of $n_w = 5$. It is striking that the QCD values obtained using our method are always higher than the ones derived using DCC, which indicates that dFC clustering would result in more reliable mind states.

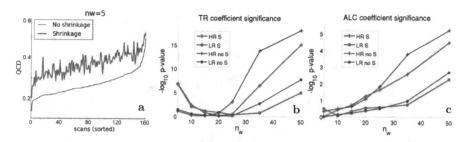

Fig. 3. (a) QCD measures obtained by applying the two approaches to all high-resolution timeseries with the effective sample size being $n_w = 5$. To ease visualization, the scans are sorted according to their QCD values under the baseline approach. (b–c) negative log p-value of the correlation between QCD and fMRI repetition time (TR) and alcohol diagnosis (ALC) for high-resolution data (HR) and low-resolution data (LR) as well as with shrinkage (S) or without covariance shrinkage (no S).

For both high resolution and low resolution clinical data sets, we also modelled the relation between QCD and explanatory variables by computing the following generative additive model

$$QCD \sim \alpha + \beta_0 sex + \beta_1 age + \beta_2 ALC + \beta_3 TR. \tag{12}$$

We then plotted in Fig. 3 the negative log of the p-value for those explanatory variables, whose correlation with QCD is significant ($p < 0.05$). Specifically, the QCDs generated by both implementations recorded significant correlations with alcohol diagnosis ALC and acquisition protocol TR. Figure 3(b) and (c) show agreement between the results obtained for low and high resolution data sets. Specifically, for small effective sample sizes the impact of the explanatory variables on the QCDs is similar for high and low resolution data. For large effective sample sizes, the effect of the explanatory variables on the QCD is always higher for the baseline approach. Assuming that the results obtained for our synthetic data set translate to real fMRI data, these results would suggest that the findings generated by the approach without covariance shrinkage, i.e., DCC, overestimate the impact of explanatory variables for large effective sample sizes.

3.5   Networks Extraction from High Resolution Data

For the final experiment, we extracted the Default Mode Network (DMN) from the dFC by slightly modifying both approaches. Specifically, we confined the computations of Eq. 2+3 to the correlations between the average BOLD signal within the Posterior Cingulate Cortex (PCC) and all the gray matter voxels in the atlas. Of the two approaches, Fig. 4 reveals that the Default Mode Networks extracted from a small number of dFC matrices by our method are closer to the typical DMN (scan average). Indeed the Spearman correlation with respect to the scan average was 0.72 on the high resolution data and 0.85 on the low resolution

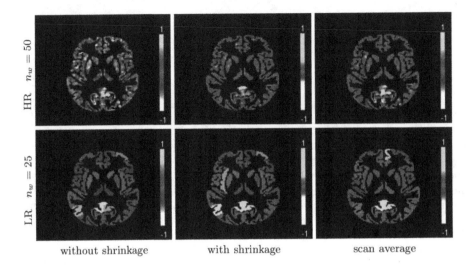

without shrinkage   with shrinkage   scan average

Fig. 4. Default mode network (DMN) extraction by averaging over 5 randomly selected dFCs estimated without (Left) and with shrinkage (Middle); Standard DMN extraction by averaging over the entire time series (scan average). On both low and high resolution data, the networks generated from the approach with shrinkage are much closer to the scan average than without shrinkage.

data for the approach with covariance shrinkage. The correlations drop down to 0.61 on the high-resolution data and 0.75 on the low resolution data for the approach without covariance shrinkage. In summary, this experiment suggests that covariance shrinkage improves functional networks extraction when only a few BOLD measurements are available for the estimation.

4   Conclusion

In this paper, we demonstrated that linear covariance shrinkage improved the estimation of dynamic resting-state fMRI connectivity, in particular when short brain connectivity states are extracted. We showed how the computation of large covariance matrices can be averted to scale the processing to full resolution fMRI scans, which suppressed the need of defining summary statistics for functional brain atlases regions. Our experiments on synthetic data demonstrate that our approach produced dFC estimates that are closer to the ground-truth. On real data, we found that our approach is better at extracting functional networks and capturing differences in rs-fMRI acquisition and diagnosis.

Acknowledgements. This research work was funded by the National Institute on Alcohol Abuse and Alcoholism (NIAAA) under the grants AA005965, AA013521, AA010723, and AA026762.

References

1. Biswal, B., Zerrin Yetkin, F., Haughton, V., Hyde, J.: Functional connectivity in the motor cortex of resting human brain using echo-planar MRI. Magn. Reson. Med. **34**(4), 537–541 (1995)
2. Chang, C., Glover, G.: Time-frequency dynamics of resting-state brain connectivity measured with fMRI. NeuroImage **50**(1), 81–98 (2010)
3. Chen, Y., Wiesel, A., Eldar, Y.C., Hero, A.O.: Shrinkage algorithms for MMSE covariance estimation. IEEE Trans. Signal Process. **58**(10), 5016–5029 (2010)
4. Hofmann, T., Schölkopf, B., Smola, A.: Kernel methods in machine learning. Ann. Stat. **36**(3), 1171–1220 (2008)
5. Ledoit, W.: Improved estimation of the covariance matrix of stock returns with an application to portfolio selection. J. Empirical Finan. **10**(5), 603–621 (2003)
6. Lindquist, M., Xu, Y., Nebel, M., Caffo, B.: Evaluating dynamic bivariate correlations in resting-state fMRI: a comparison study and a new approach. NeuroImage **101**, 531–546 (2014)
7. Liu, X., Zhang, N., Chang, C., Duyn, J.: Co-activation patterns in resting-state fMRI signals. NeuroImage **180**(Part B), 485–494 (2018)
8. Pfefferbaum, A., et al.: Accelerated aging of selective brain structures in human immunodeficiency virus infection. Neurobiol. Aging **35**(7), 1755–1768 (2014)
9. Preti, M., Bolton, T., Van De Ville, D.: The dynamic functional connectome: state-of-the-art and perspectives. NeuroImage **160**, 41–54 (2017)
10. Rohlfing, T., Zahr, N., Sullivan, E., Pfefferbaum, A.: The SRI24 multichannel atlas of normal adult human brain structure. Hum. Brain Mapp. **31**(5), 798–819 (2014)
11. Xie, H., et al.: Efficacy of different dynamic functional connectivity methods to capture cognitively relevant information. NeuroImage **188**, 502–514 (2019)

Rapid Acceleration of the Permutation Test via Transpositions

Moo K. Chung[1]([✉]), Linhui Xie[2], Shih-Gu Huang[1], Yixian Wang[1], Jingwen Yan[2], and Li Shen[3]

[1] University of Wisconsin, Madison, USA
mkchung@wisc.edu
[2] Indiana University-Purdue University Indianapolis, Indianapolis, USA
[3] University of Pennsylvania, Philadelphia, USA

Abstract. The permutation test is an often used test procedure for determining statistical significance in brain network studies. Unfortunately, generating every possible permutation for large-scale brain imaging datasets such as HCP and ADNI with hundreds of subjects is not practical. Many previous attempts at speeding up the permutation test rely on various approximation strategies such as estimating the tail distribution with known parametric distributions. In this study, we propose the novel transposition test that exploits the underlying algebraic structure of the permutation group. The method is applied to a large number of diffusion tensor images in localizing the regions of the brain network differences.

Keywords: Permutation test · Transposition test · Structural brain networks · Permutation group · Online statistics computation

1 Introduction

The permutation test is perhaps the most widely used nonparametric test procedure in sciences [8,19,21,24,27]. It is known as the exact test in statistics since the distribution of the test statistic under the null hypothesis can be exactly computed if we can calculate all the test statistics under every possible permutation. Unfortunately, generating every possible permutation for a large-sample network dataset is still extremely time consuming even for a modest sample size.

When the total number of permutations is large, various resampling techniques have been proposed to speed up the computation in the past [8,19,21,27]. In the resampling methods, only a small fraction of possible permutations is generated and the statistical significance is computed *approximately*. This approximate permutation test is the most widely used version of the permutation test. In most of brain imaging studies, 5,000–1,000,000 permutations are often used, which puts the total number of generated permutations usually less than a fraction of all possible permutations. In [27], 5,000 permutations out of possible

© Springer Nature Switzerland AG 2019
M. D. Schirmer et al. (Eds.): CNI 2019, LNCS 11848, pp. 42–53, 2019.
https://doi.org/10.1007/978-3-030-32391-2_5

$\binom{27}{12} = 17,383,860$ permutations (0.029%) were used. In [21], 1 million permutations out of $\binom{40}{20}$ possible permutations (0.0007%) were generated using a super computer. In [18], 5,000 permutations out of possible $\binom{33}{10} = 92561040$ permutations (0.005%) were used.

To remedy the computational bottleneck, the tail regions of the distributions are often estimated using the extreme value theory [11,24]. One main tool in the extreme value theory is the use of generalized Pareto distribution in approximating the tail distributions. Unfortunately, without a prior information or model fit, it is difficult to even guess the shape of tails accurately. Recently, an exact topological inference approach with quadratic run time was proposed to combinatorially enumerate every possible permutation [8,9], but the method is limited to monotone network features and not applicable to more general settings.

In this paper, we propose a novel transposition test that is motived by the permutation test. The method is based on the concept of random transpositions that sequentially update the test statistic. Unlike the traditional permutation test that takes up to a few days on a computer, our method takes less than an hour and does not require large computer memory. As an application, the method is used in determining the statistical significance of the male and female differences in a large-sample structural brain network study.

2 Preliminary

The usual statistical test setting in brain imaging is a two-sample comparison [8,19,21]. Consider two ordered sets

$$\mathbf{x} = (x_1, x_2, \cdots, x_m), \quad \mathbf{y} = (y_1, y_2, \cdots, y_n).$$

The distance between $\mathbf{x}$ and $\mathbf{y}$ is measured by test statistic $f(\mathbf{x}, \mathbf{y})$ such as t-statistic or correlations. Under the null assumption of the equivalence of $\mathbf{x}$ and $\mathbf{y}$, elements in $\mathbf{x}$ and $\mathbf{y}$ are permutable. Consider combined ordered set $\mathbf{z} = (x_1, \cdots, x_m, y_1, \cdots, y_n)$ and its all possible permutations $\mathbb{S}_{m+n}$. Note $\mathbb{S}_{m+n}$ is a symmetric group of order $m + n$ with $(m + n)!$ possible permutations [14]. Since there is an isomorphism between $\mathbf{z}$ and integer set $\{1, 2, \cdots, m + n\}$, we will interchangeably use them when appropriate [17]. Permutation $\tau \in \mathbb{S}_{m+n}$ is often denoted as

$$\tau = \begin{pmatrix} x_1 & \cdots & x_m & y_1 & \cdots & y_n \\ \tau(x_1) & \cdots & \tau(x_m) & \tau(y_1) & \cdots & \tau(y_n) \end{pmatrix}$$

using a single combined sample notation in mathematical literature [10,14].

For instance, consider a permutation of $\{1, 2, 3, 4\}$ given by $\tau(1) = 4, \tau(2) = 2, \tau(3) = 1, \tau(4) = 3$. Since there are two cycles in the permutation, τ can be written in the cyclic form as $\tau = [2][1, 4, 3]$ indicating 2 is a cycle of length 1 $(2 \to 2)$ while 1, 3, 4 are a cycle of length 3 $(1 \to 4 \to 3 \to 1)$ [14]. A cycle of length 1 is simply ignored and the permutation can be written as $\tau = [1, 4, 3]$. If another permutation is given by $\pi(1) = 1, \pi(2) = 4, \pi(3) = 3, \pi(4) = 2$, the sequential application of π to τ is written as

$$\pi \cdot \tau = [1][3][2,4] \cdot [2][1,4,3] = [2,4] \cdot [1,4,3] = [1,2,4,3].$$

Let us split the permutation $\tau(\mathbf{z})$ into two groups with m and n elements

$$\tau(\mathbf{x}) = (\tau(x_1), \cdots, \tau(x_m)), \quad \tau(\mathbf{y}) = (\tau(y_1), \cdots, \tau(y_n)).$$

For test statistic f, the *exact p*-value for testing a one sided hypothesis is then given by the fraction

$$p\text{-value} = \frac{1}{(m+n)!} \sum_{\tau \in \mathbb{S}_{m+n}} \mathcal{I}\big(f(\tau(\mathbf{x}), \tau(\mathbf{y})) > f(\mathbf{x}, \mathbf{y})\big), \tag{1}$$

where $\mathcal{I}$ is an indicator function taking value 1 if the argument is true and 0 otherwise. In various brain imaging applications, computing statistic f for each permutation has been the main computational bottleneck [8, 21].

If the test statistic f is a symmetric function such that $f(\mathbf{x}, \mathbf{y}) = f(\phi(\mathbf{x}), \psi(\mathbf{y}))$, where $\phi \in \mathbb{S}_m$ and $\psi \in \mathbb{S}_n$, then we only need to consider $\binom{m+n}{m}$ number of permutations in the denominator of (1), which reduces the number of possible permutations substantially. Still $\binom{m+n}{m}$ is an extremely large number and most computing systems including MATLAB/R cannot compute them exactly if the sample size is larger than 100 in each group. The total number of permutations when $m = n$ is given asymptotically by Stirling's formula [12]

$$\binom{2m}{m} \sim \frac{4^m}{\sqrt{\pi m}}.$$

The number of permutations *exponentially* increases as the sample size increases, and thus it is impractical to generate every possible permutation. In practice, up to hundreds of thousands of random permutations are generated using the uniform distribution on $\mathbb{S}_{m+n}$ with probability $1/\binom{m+n}{m}$.

3   Methods

Transpositions. Consider permutation π_{ij} that exchanges i-th and j-th elements between $\mathbf{x}$ and $\mathbf{y}$ and keeps all others fixed such that

$$\pi_{ij}(\mathbf{x}) = (x_1, \cdots, x_{i-1}, y_j, x_{i+1}, \cdots, x_m),$$
$$\pi_{ij}(\mathbf{y}) = (y_1, \cdots, y_{j-1}, x_i, y_{j+1}, \cdots, y_n).$$

Such a permutation is called the *transposition*, which is related to card shuffling problems [1, 2, 14]. Consider every possible sequence of transpositions applied to $\mathbf{x}$ and $\mathbf{y}$. If such sequence of transpositions covers every possible element in $\mathbb{S}_{m+n}$, we can perform the permutation test by sequentially transposing two elements at a time.

Theorem 1. *Any permutation in $\mathbb{S}_{m+n}$ can be reachable by a sequence of transpositions.*

Proof. Let $l = m + n$. Suppose $\tau \in \mathbb{S}_l$. For $x \in \{1, \cdots, l\}$, consider cycle

$$C_x = [x, \tau(x), \tau^2(x), \cdots, \tau^j(x)]$$

with $\tau^{j+1}(x) = x$ and $\tau^d(x) \neq x$ for $d \leq j$ [14]. Since we are dealing with a finite number of elements, such j always exists. If $\tau^c(x) = \tau^d(x)$ for some $c \leq d \leq j$, we have $\tau^{d-c}(x) = x$, thus all elements in the cycle C_x are distinct.

If C_x covers all the elements in $\{1, \cdots, l\}$ we proved the statement. If there is an element, say $y \in \{1, \cdots, l\}$, that is not covered by C_x, we construct a new cycle C_y. Cycles C_x and C_y must be disjoint. If not, we have $\tau^i(x) = \tau^j(y)$ and $y = \tau^{i-j}(x)$, which is in contradiction. Hence $\tau = C_x \cdot C_y$.

If $C_x \cdot C_y$ does not cover $\{1, \cdots, l\}$, we repeat the process until we exhaust all the elements in $\{1, \cdots, l\}$. Hence any permutation can be decomposed as a product of *disjoint* cycles. Then algebraic derivation can further show that cycle C_x can be decomposed as a product of 2-cycles

$$C_x = [x, \tau^j(x)] \cdot [x, \tau^{j-1}(x)] \cdots [x, \tau^2(x)] \cdot [x, \tau(x)].$$

A 2-cycle is simply a walk. Hence we proved τ is a sequence of walks.   □

From Theorem 1, any permutation can be reached by a sequence of transpositions. Thus, instead of performing uniform random sampling in $\mathbb{S}_{m+n}$, we will perform a sequence of random transpositions and compute the test statistic at each transposition. Over random transposition π_{ij}, the statistic changes from $f(\mathbf{x}, \mathbf{y})$ to $f(\pi_{ij}(\mathbf{x}), \pi_{ij}(\mathbf{y}))$. Instead of computing $f(\pi_{ij}(\mathbf{x}), \pi_{ij}(\mathbf{y}))$ directly, we will compute it from $f(\mathbf{x}, \mathbf{y})$ incrementally in *constant* run time by updating the value of $f(\mathbf{x}, \mathbf{y})$. Note, the statistics computation over transpositions is slightly different from the usual online computation where new data is added sequentially. Instead of adding the new data, the existing data is replaced.

Theorem 2. *If f is an algebraic function such as addition, subtraction, multiplication, division and integer exponents, there exists a function g such that*

$$f(\pi_{ij}(\mathbf{x}), \pi_{ij}(\mathbf{y})) = g(f(\mathbf{x}, \mathbf{y}), x_i, y_j), \tag{2}$$

where the computational complexity of g is constant.

The lengthy proof involves explicitly constructing an iterative formula for each algebraic operation so it will not be shown here. Often used statistics such as t-statistic and F-statistic are all algebraic functions. If we take computation involving *fractional exponents* as constant run time as well, then a much wider class of statistics such as correlations can all have constant run time. In this study, we will explicitly construct the t-statistic over transpositions that runs in constant run time. From this construction, it should be obvious that Theorem 2 should be applicable to other test statistics.

t-Statistic over a Transposition. Two sample t-statistic is a function of symmetric functions involving the mean and variance of $\mathbf{x}$ and $\mathbf{y}$. If we have the symmetric functions

$$\nu(\mathbf{x}) = \sum_{j=1}^{m} x_j, \quad \omega(\mathbf{x}) = \sum_{j=1}^{m} \left(x_j - \frac{\nu(\mathbf{x})}{m}\right)^2,$$

the sample mean and variance of $\mathbf{x}$ are given by $\nu(\mathbf{x})/m$ and $\omega(\mathbf{x})/(m-1)$. We will determine how ν and ω change over a transposition.

Theorem 3. *Functions ν and ω are updated over transposition π_{ij} as*

$$\nu(\pi_{ij}(\mathbf{x})) = \nu(\mathbf{x}) - x_i + y_j$$
$$\omega(\pi_{ij}(\mathbf{x})) = \omega(\mathbf{x}) - x_i^2 + y_j^2 + \frac{\nu(\mathbf{x})^2 - \nu(\pi_{ij}(\mathbf{x}))^2}{m}.$$

Proof. The algebraic derivation follows applying the online computations of updating ν and ω twice. Suppose new data a and b is added to $\mathbf{x}' = (x_1, \cdots, x_{m-1})$ such that $\mathbf{x}_a = (\mathbf{x}', a)$ and $\mathbf{x}_b = (\mathbf{x}', b)$. Then we have

$$\nu(\mathbf{x}_b) = \nu(\mathbf{x}_a) - a + b.$$

Since ν is symmetric, by identifying $a = x_i$ and $b = y_j$, we obtain $\nu(\pi_{ij}(\mathbf{x})) = \nu(\mathbf{x}) - x_i + y_j$. An algebraic derivation can show that

$$\omega(\mathbf{x}_a) = \sum_{j=1}^{m-1} x_j^2 + a^2 - \frac{\nu(\mathbf{x}_a)^2}{m}, \quad \omega(\mathbf{x}_b) = \sum_{j=1}^{m-1} x_j^2 + b^2 - \frac{\nu(\mathbf{x}_b)^2}{m}. \quad (3)$$

From (3), we obtain $\omega(\mathbf{x}_b) = \omega(\mathbf{x}_a) - a^2 + b^2 + \frac{\nu(\mathbf{x}_a)^2 - \nu(\mathbf{x}_b)^2}{m}$ and the result follows. $\square$

From Theorem 3, the two-sample t-statistic over a transposition is then computed as follows.

$$T(\pi_{ij}(\mathbf{x}), \pi_{ij}(\mathbf{y})) = \frac{\nu(\pi_{ij}(\mathbf{x}))/m - \nu(\pi_{ij}(\mathbf{y}))/n}{\sqrt{\frac{\omega(\pi_{ij}(\mathbf{x})) + \omega(\pi_{ij}(\mathbf{y}))}{m+n-2}\left(\frac{1}{m} + \frac{1}{n}\right)}}.$$

Computing two-sample t-statistic with m and n samples directly requires computing the sample means, which is m and n algebraic operations each. Then we need to compute the sample variances and pool them together, which requires $3m + 2$ and $3n + 2$ operations. Combining the numerator and denominator in t-statistic takes 16 operations. Thus, it takes total $3(m + n) + 20$ operations to compute the t-statistic per permutation. In comparison, it only takes 35 operations to computer t-statistic per transposition. In the case of $m = n = 200$, the proposed method can generate 1220 times more permutations compared to the standard permutation test.

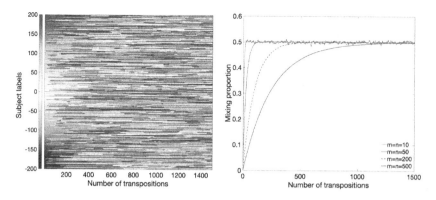

Fig. 1. Left: Mixing of subject labels over transpositions. Right: The estimated mixing proportion based on the average of 1000 simulations.

Reducing Mixing Time. Given $m = n$ elements in each group, the standard permutation test mixes half of elements in one group to the other. Thus, the mixing proportion is 0.5 on average. On the other, the transposition method mixes one element at a time, so the mixing is slow but it rapidly catches up. The rate of mixing can be formally measured by the *mixing time*, which is defined as the time until the transpositions are close to the uniform distribution in $\mathbb{S}_{m+n}$ in the variation distance sense [2,5]. Even though the transposition method does not guarantee the uniform distribution in $\mathbb{S}_{m+n}$ in the early stage of transpositions, the method converges to the uniform distribution quickly in $O((m+n)log(m+n))$ time [2,5]. This is demonstrated in Fig. 1.

Figure 1-left displays how the subject labels change over transpositions based on the sample sizes $m = n = 200$. The first group is indexed between 1 and 200 while the second group is indexed between -1 and -200. At each transposition, only two subjects are swapped. As the number of transpositions increases, subject labels rapidly mix up. Figure 1-right shows that how the mixing proportion converges to 0.5 based on the average of 1000 simulations. On average, about 1000 transpositions are enough to mix all the elements in the two groups uniformly. For far smaller sample sizes, to which most brain imaging studies belong, few hundreds transpositions are more than enough to mix the groups evenly.

To increase the rate of mixing further, we did not start with the original data **x** and **y** but started with a random permutation of **x** and **y**. This has the effect of starting with a completely mixed initial starting data. Then sequentially applied 5,000 random transpositions. This process is iteratively repeated. Thus, for every 5,000 random transpositions, one random permutation is intermixed. In real data, this process is repeated 10,000 times to generate 50 million random transpositions, which are intermixed with 10,000 permutations.

Multiple Comparisons. So far we have shown how test statistics change over transpositions. We now show how the multiple comparison corrected p-values are affected over transpositions. Suppose $\mathbf{x}(q)$ and $\mathbf{y}(q)$ are functional data on edge q in a brain network $\mathcal{M}$. Given statistic map $h(q) = f(\mathbf{x}(q), \mathbf{y}(q))$ at the edge level, the hypotheses of interests are given by

$$H_0 : h(q) = 0 \text{ for all } q \in \mathcal{M} \quad vs. \quad H_1 : h(q) > 0 \text{ for some } q \in \mathcal{M}. \quad (4)$$

Once the iterative algorithm for computing the test statistic is identified, the p-value for pointwise inference at each *fixed* q can be computed iteratively. At the k-th random transposition, the uncorrected p-value is given as p_k. Then p_{k+1} is computed from iterative formula

$$(k+1)p_{k+1} = kp_k + \mathcal{I}\Big(f(\pi_{ij}(\mathbf{x}), \pi_{ij}(\mathbf{y})) \geq f(\mathbf{x}, \mathbf{y}) \Big), \quad (5)$$

where π_{ij} changes over random transpositions. Note the p-value for multiple comparisons over all q is given by

$$p\text{-value} = P\Big(\bigcup_{q \in \mathcal{M}} \{h(q) > c\} \Big) = 1 - P\Big(\bigcap_{q \in \mathcal{M}} \{h(q) \leq c\} \Big)$$
$$= 1 - P\Big(\sup_{q \in \mathcal{M}} h(q) \leq c \Big)$$

for some threshold c [25]. Thus, for multiple comparisons, the formula (5) changes to

$$(k+1)p_{k+1} = kp_k + \mathcal{I}\Big(\sup_{q \in \mathcal{M}} h(\pi_{ij}(\mathbf{x}(q)), \pi_{ij}(\mathbf{y}(q))) \geq \sup_{q \in \mathcal{M}} h(\mathbf{x}(q), \mathbf{y}(q)) \Big).$$

For alternate hypothesis $H_1 : h(q) < 0$, a similar algorithm can be used for test statistic $\inf_{q \in \mathcal{M}} h(q)$.

Validation. The real data does not have the ground truth. Thus, we compared the proposed transposition method against the standard permutation test in random simulations with the ground truth. We simulated $x_1, \cdots, x_m \sim N(0, 1)$, standard normal distribution, and $y_1, \cdots, y_n \sim N(0.1, 1)$, which provides the *ground truth* in computing t-statistic and p-value. The use of t-statistic is the standard validation framework in many previous permutation test studies [6, 13, 19, 24]. The simulations were independently performed 100 times and their average was reported here. In both Simulations 1 and 2 below, we used the same model but different sample sizes.

Simulation 1 (small sample size). We used $m = n = 10$. There are exactly $\binom{20}{10} = 184756$ total permutations. The sample sizes are too small to differentiate the group difference. We obtained the t-statistic value of 0.0533, which corresponds to the exact p-value of 0.479 (Fig. 2a green line). We performed the standard permutation test with up to 10,000 permutations, which took 0.0926 s on average on a desktop computer. Within the same run time, we were able

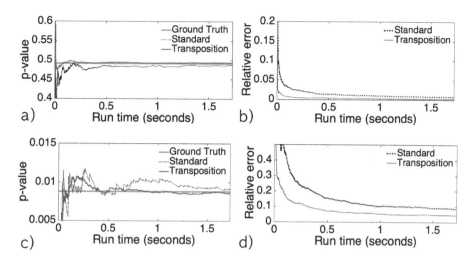

Fig. 2. (a, c) One representative simulation study showing faster convergence of the transposition method. The Gaussian distribution provides the exact ground truth. (b, d) The average relative error against the ground truth. The average of 100 independent simulations was plotted. (Color figure online)

to generate more than 1,220,000 transpositions. The transposition method uniformly converged faster than the standard permutation test due to 122 times more permutations the transposition method generated (Fig. 2b). The relative errors of the transposition method are about half the size of the standard method in most run time.

Simulation 2 (large sample size). We used $m = n = 100$. The sample sizes are big enough to differentiate the group difference. We obtained the t-statistic value of 2.39 and corresponding p-value of 0.0088, which are taken as the ground truth (Fig. 2c green line). We performed the standard permutation test with up to 1 million permutations, which took 173 s per simulation on average. With the same run time, the transposition was sequentially done about 125 million times. The transposition method uniformly converged faster than the standard permutation test through the whole run time (Fig. 2d). The performance results did not change much even if we performed more permutations over longer durations with different simulation parameters.

The computer code for performing the transposition test in the above simulation study is available at http://www.stat.wisc.edu/~mchung/transpositions. The MATLAB code is written in such a way that it accepts vector data. For brain connectivity matrices that are symmetric, vectorizing the upper triangle entries of connectivity matrices is necessary to reduce the computational time.

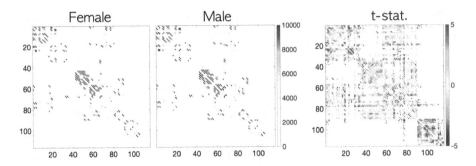

Fig. 3. Average connectivity matrices of females (left) and males (right) between 116 AAL parcellations. The two-sample t-statistic result (female − male). Females have more structural connections between brain regions than males.

4 Application

Subjects and Preprocessing. Diffusion weighted imaging (DWI) data of 202 female and 154 male subjects (ages 29.2 ± 3.4) were obtained from the Human Connectome Project (HCP) [16]. The fiber orientation distribution functions were estimated and apparent fiber densities were exploited to produce reliable WM/GM/CSF volume maps [7,16]. Subsequently, random seeds on the basis of the voxel were selected to generate about initial 10 million streamlines per subject with the maximum fiber tract length at 250 mm and FA larger than 0.06 using MRtrix3 (http://www.mrtrix.org) [22,26]. The Spherical-Deconvolution Informed Filtering of Tractograms (SIFT2) technique making use of complete streamlines was subsequently applied to generate more biologically accurate brain connectivity, which results in about 1 million tracts per subject [20]. Non-linear diffeomorphic registration between subject images to the template was performed using ANTS [3,4]. Automated Anatomical Labeling (AAL) was used to parcellate the brain into 116 regions [23]. The subject-level connectivity matrices were constructed by counting the number of tracts connecting between brain regions.

Transposition Test. We are interested in testing and localizing the female and male differences in structural connectivity. Figure 3 displays the result of group averages and the two sample t-statistic (female − male). Females have more structural connections between brain regions than males. Since the tract counts between brain regions do not follow normal distributions, assumption-free non-parametric procedures such as the permutation test are needed to determine the statistical significance of t-statistic accurately. We use the proposed transposition test by sequentially generating 50 million transpositions for 40 min on a desktop computer. For multiple comparisons correction, we counted the fraction of transpositions where the maximum t-statistic value over the whole connections is above the observed maximum t-statistic value. Any t-statistic value below

-4.05 and above 3.96 is statistically significant at 0.05 (Fig. 4). The statistically significant connections are shown in Fig. 5.

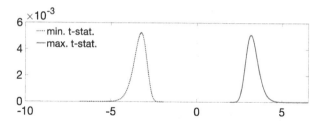

Fig. 4. The empirical distributions of minimum t-statistic (dotted blue) and maximum t-statistic (solid red), which do not follow well known statistical distributions. The proposed method is used to compute the multiple comparisons corrected p-value. (Color figure online)

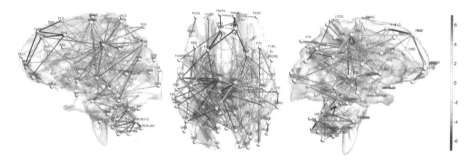

Fig. 5. t-statistic map (female $-$ male). Only the connections that are statistically significant (thresholded at -4.05 and 3.96) after multiple comparisons correction at 0.05 are shown. Females have more connections in most parts of the brain while males are more connected in the frontal regions of the brain.

Sex Difference in Connectivity. Females have far more connections in most parts of the brain while males have more connections in the frontal regions of the brain. Females also have more bilateral connections between the hemispheres. This indicates that females use the both sides of the brain while males only use one side of the brain. Females have more connections in the limbic structures that regulate emotions. Males have more connections between the back and frontal regions. Our findings are consistent with the previous structural connectivity study [15].

5   Discussion

Although we did not show here, it is also possible to construct incremental procedures for computing other test statistics such as F-statistic and Hotelling's T^2 statistic over random transpositions. These problems are left as future studies.

Compared to other approximate strategies for the permutation test, the proposed method is assumption and model free. The tail approximation method in [24] has parametric model assumptions to fit the tail regions, so the tail of the distribution needs to follow some specific pattern. On the other hand, the proposed method has no assumption on the distribution other than permutability between the groups and offers far more flexibility than [24].

We did not perform the comparisons between the methods in the real data since there is no ground truth. Thus the comparisons were done on the simulation, where the ground truths are exactly given.

Acknowledgements. This work was supported by NIH grant R01 EB022856, R01 EB022574 and NSF IIS 1837964. We would like to thank Jean-Baptiste Poline of McGill University, John Kornak of University of California - San Fransisco and Michale A. Newton of University of Wisconsin - Madison for valuable comments and discussions on the mixing time of the transposition test.

References

1. Aldous, D.: Random walks on finite groups and rapidly mixing Markov chains. In: Azéma, J., Yor, M. (eds.) Séminaire de Probabilités XVII 1981/82. LNM, vol. 986, pp. 243–297. Springer, Heidelberg (1983). https://doi.org/10.1007/BFb0068322
2. Aldous, D., Diaconis, P.: Shuffling cards and stopping times. Am. Math. Monthly **93**, 333–348 (1986)
3. Avants, B., Epstein, C., Grossman, M., Gee, J.: Symmetric diffeomorphic image registration with cross-correlation: evaluating automated labeling of elderly and neurodegenerative brain. Med. Image Anal. **12**, 26–41 (2008)
4. Avants, B., Tustison, N., Song, G., Cook, P., Klein, A., Gee, J.: A reproducible evaluation of ANTs similarity metric performance in brain image registration. NeuroImage **54**, 2033–2044 (2011)
5. Berestycki, N., Schramm, O., Zeitouni, O.: Mixing times for random k-cycles and coalescence-fragmentation chains. Ann. Probab. **39**, 1815–1843 (2011)
6. Bullmore, E., Suckling, J., Overmeyer, S., Rabe-Hesketh, S., Taylor, E., Brammer, M.: Global, voxel, and cluster tests, by theory and permutation, for difference between two groups of structural MR images of the brain. IEEE Trans. Med. Imaging **18**, 32–42 (1999)
7. Christiaens, D., Reisert, M., Dhollander, T., Sunaert, S., Suetens, P., Maes, F.: Global tractography of multi-shell diffusion-weighted imaging data using a multi-tissue model. NeuroImage **123**, 89–101 (2015)
8. Chung, M.K., Luo, Z., Leow, A.D., Alexander, A.L., Davidson, R.J., Hill Goldsmith, H.: Exact combinatorial inference for brain images. In: Frangi, A.F., Schnabel, J.A., Davatzikos, C., Alberola-López, C., Fichtinger, G. (eds.) MICCAI 2018. LNCS, vol. 11070, pp. 629–637. Springer, Cham (2018). https://doi.org/10.1007/978-3-030-00928-1_71
9. Chung, M.K., Villalta-Gil, V., Lee, H., Rathouz, P.J., Lahey, B.B., Zald, D.H.: Exact topological inference for paired brain networks via persistent homology. In: Niethammer, M., et al. (eds.) IPMI 2017. LNCS, vol. 10265, pp. 299–310. Springer, Cham (2017). https://doi.org/10.1007/978-3-319-59050-9_24
10. Dummit, D., Foote, R.: Abstract Algebra. Wiley, Hoboken (2004)

11. Embrechts, P., Resnick, S., Samorodnitsky, G.: Extreme value theory as a risk management tool. North Am. Actuarial J. **3**, 30–41 (1999)
12. Feller, W.: An Introduction to Probability Theory and its Applications, vol. 2. Wiley, Hoboken (2008)
13. Hayasaka, S., Phan, K.L., Liberzon, I., Worsley, K.J., Nichols, T.E.: Nonstationary cluster-size inference with random field and permutation methods. Neuroimage **22**, 676–687 (2004)
14. Hungerford, T.: Algebra. Springer, New York (1980)
15. Ingalhalikar, M., et al.: Sex differences in the structural connectome of the human brain. Proc. Nat. Acad. Sci. **111**, 823–828 (2014)
16. Jeurissen, B., Tournier, J.D., Dhollander, T., Connelly, A., Sijbers, J.: Multi-tissue constrained spherical deconvolution for improved analysis of multi-shell diffusion MRI data. NeuroImage **103**, 411–426 (2014)
17. Kondor, R., Howard, A., Jebara, T.: Multi-object tracking with representations of the symmetric group. In: International Conference on Artificial Intelligence and Statistics (AISTATS), vol. 1, p. 5 (2007)
18. Lee, H., Kang, H., Chung, M., Lim, S., Kim, B.N., Lee, D.: Integrated multimodal network approach to PET and MRI based on multidimensional persistent homology. Hum. Brain Mapp. **38**, 1387–1402 (2017)
19. Nichols, T., Holmes, A.: Nonparametric permutation tests for functional neuroimaging: a primer with examples. Hum. Brain Mapp. **15**, 1–25 (2002)
20. Smith, R., Tournier, J.D., Calamante, F., Connelly, A.: SIFT2: enabling dense quantitative assessment of brain white matter connectivity using streamlines tractography. NeuroImage **119**, 338–351 (2015)
21. Thompson, P., et al.: Genetic influences on brain structure. Nat. Neurosci. **4**, 1253–1258 (2001)
22. Tournier, J., Calamante, F., Connelly, A., et al.: MRtrix: diffusion tractography in crossing fiber regions. Int. J. Imaging Syst. Technol. **22**, 53–66 (2012)
23. Tzourio-Mazoyer, N., et al.: Automated anatomical labeling of activations in spm using a macroscopic anatomical parcellation of the MNI MRI single-subject brain. NeuroImage **15**, 273–289 (2002)
24. Winkler, A., Ridgway, G., Douaud, G., Nichols, T., Smith, S.: Faster permutation inference in brain imaging. NeuroImage **141**, 502–516 (2016)
25. Worsley, K., Marrett, S., Neelin, P., Vandal, A., Friston, K., Evans, A.: A unified statistical approach for determining significant signals in images of cerebral activation. Hum. Brain Mapp. **4**, 58–73 (1996)
26. Xie, L., et al.: Heritability estimation of reliable connectomic features. In: Wu, G., Rekik, I., Schirmer, M.D., Chung, A.W., Munsell, B. (eds.) CNI 2018. LNCS, vol. 11083, pp. 58–66. Springer, Cham (2018). https://doi.org/10.1007/978-3-030-00755-3_7
27. Zalesky, A., et al.: Whole-brain anatomical networks: does the choice of nodes matter? NeuroImage **50**, 970–983 (2010)

Heat Kernels with Functional Connectomes Reveal Atypical Energy Transport in Peripheral Subnetworks in Autism

Markus D. Schirmer[1,2] and Ai Wern Chung[3]

[1] Stroke Division and Massachusetts General Hospital,
J. Philip Kistler Stroke Research Center, Harvard Medical School, Boston, USA
[2] Department of Population Health Sciences,
German Centre for Neurodegenerative Diseases (DZNE), Bonn, Germany
[3] Fetal-Neonatal Neuroimaging and Developmental Science Center, Boston
Children's Hospital, Harvard Medical School, Boston, MA, USA
AiWern.Chung@childrens.harvard.edu

Abstract. Autism is increasing in prevalence and is a neurodevelopmental disorder characterised by impairments in communication skills and social behaviour. Connectomes enable a systems-level representation of the brain with recent interests in understanding the distributed nature of higher order cognitive function using modules or subnetworks. By dividing the connectome according to a central component of the brain critical for its function (it's hub), we investigate network organisation in autism from hub through to peripheral subnetworks. We complement this analysis by extracting features of energy transport computed from heat kernels fitted with increasing time steps. This heat kernel framework is advantageous as it can capture the energy transported in all direct and indirect pathways between pair-wise regions over 'time', with features that have correspondence to small-world properties. We apply our framework to resting-state functional MRI connectomes from a large, publically available autism dataset, ABIDE. We show that energy propagating through the brain over time are different between subnetworks, and that heat kernel features significantly differ between autism and controls. Furthermore, the hub was functionally preserved and similar to controls, however, increasing statistical significance between groups was found in increasingly peripheral subnetworks. Our results support the increasing opinion of non-hub regions playing an important role in functional organisation. This work shows that analysing autism by subnetworks with the heat kernel reflects the atypical activations in peripheral regions as alterations in energy dispersion and may provide useful features towards understanding the distributed impact of this disorder on the functional connectome.

Keywords: Connectome · Functional network · Heat kernel ·
Diffusion equation · Subnetworks · Hubs · Autism

© Springer Nature Switzerland AG 2019
M. D. Schirmer et al. (Eds.): CNI 2019, LNCS 11848, pp. 54–63, 2019.
https://doi.org/10.1007/978-3-030-32391-2_6

1   Introduction

Autism spectrum disorder is a neurodevelopmental condition estimated to affect 1 in 59 children in the US [2]. It is characterised by atypical social behaviour and sensory processing, with deficits in high-level cognitive function and mental flexibility [16,22]. It has also been increasingly suggested that the neural bases of autism may not be explained by specific regions, but by aberrant connectivity within and between functional modules [16,19].

Thus strategies to interrogate brain connectivity in autism have evolved from specific tract analysis to connectomes. This connectome approach recognises the distributed nature of higher order cognitive functions. Recent approaches have been to identify modules or subsets of regions that are most critical for efficient network function [18,26,27] and which exhibit specialisation for specific processes [11]. For one such approach, inter-connected brain regions of high functional or structural connectivity are considered to form a collection of core "hubs", a subnetwork that is essential for efficient cognitive function. Such hubs are the first to develop and present at birth, with strong similarity to adults [12,14]. Other regions that form later during development around the hubs constitute the "feeder" and "seeder" subnetworks [15,24]. Grouping connections into subnetworks allows interrogation of the relative importance of each subnetwork in characterising a disorder, such as autism, providing information on changes in the fundamental core of a network compared to secondary maladaptive changes that potentially give rise to cognitive dysfunction [3,4,28,30].

Studies using subnetworks often compare measures of density or connectivity strength [8,30], or traditional network metrics [3,24]. With brain function potentially being supported by coordinated activity between different functional modules, recent methods have sought to capture these dynamic processes [13,16]. Here, we propose to use heat kernels, a diffusion model, on resting-state functional MRI (rs-fMRI) data to extract features of energy transport [5,6]. The heat kernel describes the effect of applying a heat source to a network and observing the diffusion process over 'time'. It encodes the distribution of energy over a network and characterises the underlying structure of the graph [7,31]. Heat kernels have been applied to connectomes to investigate atrophy patterns in Alzheimer's [21], mappings between functional and structural connectomes [1] and for predicting motor outcome in preterm infants [6].

In this work, we present an edge-centric analysis where energy propagation features are computed from rs-fMRI hub-stratified subnetworks. These heat kernel features are then compared between a large, multi-centre cohort of autism and control subjects. By combining a dynamic network model with hub analysis, our aim is to better understand the vulnerability of and interplay between subnetworks in autism.

2   Materials and Methods

2.1   Subjects and rs-FMRI Data Preprocessing

We used rs-fMRI data from the Autism Brain Imaging Data Exchange (ABIDE) initiative [10], comprising of typically-developing controls (n = 440) and subjects with autism (n = 379). Data were preprocessed with the ABIDE Connectome Computation System pipeline. In brief, preprocessing steps included: removal of spikes, with slice timing and motion correction, removal of mean CSF and white matter signals, and detrending of linear and quadratic drifts. Band-pass temporal filtering (0.01–0.1 Hz) was applied after the above nuisance variable regressions. The global mean signal was not regressed from the data. rs-fMRI data were registered to the MNI template and signals averaged into regions according to the AAL atlas. The pre-processed timeseries was demeaned, and a covariance matrix [29] was computed for each subject. Omitting brainstem and cerebellar regions resulted in a 90 × 90 connectivity matrix for each subject.

2.2   Group Connectomes, Hub Organisation and Subnetworks

Group Connectome. A group-averaged network was computed from only the control population. The absolute of the connectivity matrix was taken and thresholded to retain values greater than 0.05 to remove spurious associations. A binarised, group-average adjacency matrix, W_{group}, was then computed by retaining edges in at least 75% of the control group.

Hub Organisation and Defining Subnetworks. Hub regions were identified from W_{group} by selecting the top ten nodes with the greatest strength [8]. Network edges are then labelled into subnetworks based on their connection to the hub nodes [24]: *Hub subnetwork* - Contains edges connecting two hub nodes; *Feeder subnetwork* - are edges connecting hub to non-hub nodes; and *Seeder subnetwork* - have edges connecting two non-hub nodes. We include a fourth *'Non-edge subnetwork'*, comprising of the remaining entries in the network which do not have an actual connection. These four subnetworks form the 'regions-of-interest' to group the edge-based, heat kernel features for analysis.

2.3   Computing Heat Kernels and Their Features

The following computations are performed on a network W for each subject (in both control and patient groups) found by multiplying their respective covariance matrix with W_{group}.

Graph Notation. $G = (V, E)$ where V is the set of $|V|$ nodes on which a graph is defined and $E \subseteq V \times V$ the corresponding set of edges. A subject's weighted matrix, W, is defined as $W(u, v) = w_{uv}$ where w_{uv} is the corresponding edge strength. A diagonal strength matrix, D, is defined as $D(u, u) = deg(u) = \sum_{v \in V} w_{uv}$. The Laplacian of G is $\mathcal{L} = D - W$ and the normalised Laplacian is given by $\hat{\mathcal{L}} = D^{-1/2} \mathcal{L} D^{-1/2}$.

Heat Kernel Features. The heat kernel, $H(t)$, is the fundamental solution to the standard, partial differential equation of a diffusion process,

$$\frac{\partial H(t)}{\partial t} = -\hat{\mathcal{L}}H(t), \tag{1}$$

and can be computed analytically,

$$H(t) = \exp(-t\hat{\mathcal{L}}). \tag{2}$$

$H(t)$ describes the flow of energy through G at time t where the rate of flow is governed by $\hat{\mathcal{L}}$ calculated from W. $H(t)$ is a symmetric $|V| \ times |V|$ matrix where the entry $H_{u,v}(t)$ represents the amount of heat transfer between nodes u and v after time t.

Based on heat kernels computed from Eq. 2 for a range of t, several features can be extracted for each entry in H to represent the dynamic properties of the network [3]. One measure is the intrinsic time constant, t_c, which is the time when the *relative* change in heat transfer has dropped below a given percentage. The $t_c(u,v)$ between nodes u and v for percentage threshold s is

$$t_c(u,v) = t_{max} : \left|\frac{H_{u,v}(t+\Delta t) - H_{u,v}(t)}{H_{u,v}(t)}\right|^{t_2}_{t_1} < s, \tag{3}$$

where Δt is a time step within the range of $t_1 \leq t \leq t_2$. The next set of measures are the maximal energy passed between two regions (maximal value across all heat kernels)

$$h_{peak}(u,v) = \max |H_{u,t}(t)|^{t_2}_{t_1}, \tag{4}$$

and the time that h_{peak} occurs

$$t_{peak}(u,v) = t : h_{peak}(u,v). \tag{5}$$

The last set of features represent the maximal *difference* in energy transferred between two regions,

$$h'_{peak}(u,v) = \max |H_{u,v}(t+\Delta t) - H_{u,v}(t)|^{t_2}_{t_1} \tag{6}$$

and the time that h'_{peak} occurs

$$t'_{peak}(u,v) = t : h'_{peak}(u,v). \tag{7}$$

2.4   Experimental Design

For each subject, 1500 heat kernels were computed from W for $t = [0.00, 0.01, \ldots 15.0]$. t_c was calculated for a range of thresholds $s = [2, 3, 4, 5\%]$. This results in a total of eight features for every edge in E. For each feature, the mean is calculated from edges belonging to each subnetwork, yielding a final 32 measures for each subject (number of features × number of subnetworks). Group differences of these 32 measures are tested for using independent t-tests. Multiple comparisons was accounted for with a Bonferroni corrected significance threshold of $p < 0.05/32 = 0.00156$.

3   Results

Table 1 is an overview of subject demographics. Ages were not significantly different between groups (independent t-test, $p = 0.61$).

Table 1. Demographics of subjects from ABIDE

Heading level	Controls	Patients
Number of subjects	440	379
Age (years, mean $\pm$ std)	16.27 $\pm$ 6.74	16.53 $\pm$ 7.54
Age (years, range)	6.47–56.2	7.0–58.0

Regions identified as hub nodes are listed in Table 2. These regions were predominantly deep grey matter structures and have found to be key hub regions (e.g. insular, superior medial frontal, supramarginal gyrus) elsewhere [20,23].

Table 2. Identified hub regions in controls.

Regions
Superior medial frontal - Left
Insular - Left
Insular - Right
Putamen - Left
Putamen - Right
Supramarginal gyrus - Right
Rolandic operculum - Left
Rolandic operculum - Right

Figure 1 plots the amount of energy in heat kernels with time, averaged by subnetwork, for each group. Specifically, it plots the heat kernel value, $H_{u,v}(t)$, averaged across all edges within a subnetwork, versus t. The slope and shape of each curve varies depending on the subnetwork. The non-edge subnetwork transports the least amount of energy over time, and the remaining subnetworks all exhibit a peak where energy is maximal at different t. There is also a consistent difference between groups over time after the peak - patients have lower heat kernel values than controls in the hub and feeder subnetworks, and the reverse can be observed in the seeder subnetwork.

Figure 2 plots each of the eight averaged heat kernel features, for all subnetworks and groups, and Table 3 shows features' mean (standard deviation) values and statistical significance levels between groups. There are no significant group

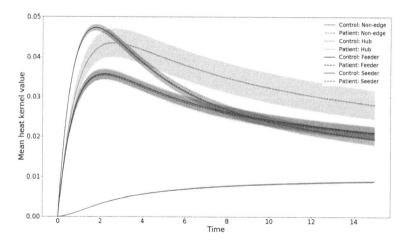

Fig. 1. Plots of mean values in the heat kernel matrix by subnetwork, with increasing time for all subjects in each group. Shaded areas represent standard deviations.

differences for all measures in the hub subnetwork, however, the more peripheral the subnetwork is to the hub, the greater the number of significant group differences, and the greater their statistical significance (Table 3). This is most apparent in t_c, irrespective of the threshold s used. Furthermore, the autism group has greater t_c than controls in all non-hub subnetworks. This trend can be similarly seen for t_{peak} and t'_{peak}.

4   Discussion

In this work, we presented an rs-fMRI subnetwork analysis comparing features of energy propagation between autism and controls. More specifically, we investigated how heat kernel derived measures differ between groups in the central functional core of the brain and its peripheral subnetworks. We found no significant difference in energy transport in hub regions between groups. However, peripheral subnetworks differed significantly, with important properties of change in energy transport occurring at later time points in autism when compared to controls.

Combining hub-stratified subnetwork analysis with the above heat kernel framework is a complementary strategy to further our understanding of brain topology. The strategic importance of hubs for information transport makes them potentially vulnerable and thus sensitive to disease [9,18,26,28]. Whereas alterations in the feeder and seeder subnetworks have been viewed as potential secondary adaptations to disease or injury [3,4,30]. However, analysis treating subnetworks as separate, standalone, entities with their own topology may be unrealistic given the highly integrative nature of the brain. Heat kernels provide a means to incorporate information from the entire network, even when computing

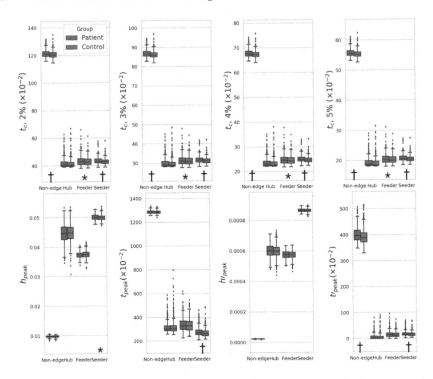

Fig. 2. Boxplots of mean heat kernel features by subnetwork, for controls versus autism. * denotes group differences at $p < 0.05$ and † for statistically significant group differences corrected for multiple comparisons at $p < 0.00156$.

edge-wise measures. This is because each element in H represents energy transport through all possible pathways that connect any two regions. It is because of this that the non-edge subnetwork possesses heat information, and has the lowest heat kernel value. The non-edge subnetwork is also highly indicative of a network's capacity (its small world propensity) for efficient energy propagation when using heat kernels [6], explaining the greatest significance between groups of all subnetworks tested in our analysis.

This combined framework revealed a number of interesting findings. The lack of significant group differences in the hub subnetwork suggests a relatively preserved functional core in autism. Subnetwork analysis in other pathologies have identified core regions to remain similar to controls, whereas peripheral regions demonstrated greater differences [3,30]. While it has been suggested that the core is stable in order to support and allow peripheral regions greater flexibility to meet functional demands [11], others have found atypical functional activation in the core in the ABIDE cohort against controls [16,17]. Differences in reported hub connectivity in autism may be attributed to not only the different methodologies used, but also because of the many ways to rate a node's importance to be labelled as a hub.

Table 3. Mean (stdev) heat kernel features by subnetwork for Control and autism groups. Statistically significant group differences are in bold denoted by $*p < 0.05$, $\dagger p < 0.00156$ (Bonferroni-corrected threshold), $\ddagger p < 0.0001$.

Feature	Non-edge		Hub		Feeder		Seeder	
	Control	Autism	Control	Autism	Control	Autism	Control	Autism
t_c, 2%	**1.207**	**1.215‡**	0.417	0.418	**0.43**	**0.436***	0.429	0.436†
	(0.027)	**(0.029)**	(0.037)	(0.035)	**(0.03)**	**(0.036)**	0.023	(0.026)
t_c, 3%	**0.864**	**0.869‡**	0.3	0.301	**0.312**	**0.316***	0.313	0.317†
	(0.019)	**(0.021)**	(0.026)	(0.025)	**(0.022)**	**(0.026)**	(0.016)	(0.019)
t_c, 4%	**0.674**	**0.679‡**	0.235	0.235	**0.245**	**0.248***	0.247	0.250†
	(0.015)	**(0.016)**	(0.021)	(0.02)	**(0.017)**	**(0.02)**	(0.013)	(0.015)
t_c, 5%	**0.554**	**0.558‡**	0.193	0.193	**0.203**	**0.205***	0.204	0.207†
	(0.012)	**(0.014)**	(0.017)	(0.017)	**(0.014)**	**(0.017)**	(0.01)	(0.012)
h_{peak}	0.976	0.974	4.473	4.465	3.747	3.733	**4.987**	**5.005***
	(0.032)	(0.03)	(0.299)	(0.303)	(0.119)	(0.107)	**(0.101)**	**(0.099)**
t_{peak}	12.84	12.856	3.213	3.159	3.333	3.36	**2.679**	**2.760†**
	(0.133)	(0.142)	(0.63)	(0.454)	(0.547)	(0.572)	**(0.339)**	**(0.385)**
h'_{peak}	0.235	0.235	5.988	5.985	5.729	5.733	8.629	8.641
	(0.007)	(0.007)	(0.43)	(0.424)	(0.246)	(0.248)	(0.113)	(0.108)
t'_{peak}	**3.937**	**4.018†**	0.067	0.069	0.142	0.157	**0.155**	**0.175†**
	(0.294)	**(0.306)**	(0.119)	(0.112)	(0.099)	(0.119)	**(0.079)**	**(0.091)**

In terms of peripheral subnetworks, the recruitment of more seeder regions has been found in long, indirect functional pathways in autism than in controls [16]. The authors suggest this may be indicative of diminished segregation between core and peripheral subnetworks. It is thus interesting that the more peripheral the subnetwork, the greater the statistical significance of our group differences measured by heat kernel features. This result in non-hub regions was also accompanied by greater values in autism than in controls for all time related features. Suggesting that while heat kernel values have a similar profile in both subject groups (Fig. 1), the extracted features quantifying properties of these curves appear at a later t in autism. As it stands, we cannot ascertain what these changes in the heat kernel features and their timings represent, but taken all together, our results suggest greater involvement from peripheral, rather than hub regions in autism.

There are limitations to our study, one of which is the broad age range in the ABIDE dataset. It is important to understand the potential impact of age on our results, particularly as the age range encompasses a period of great neurodevelopmental change from childhood through to early adulthood. Another is our use of node strength to identify hubs when a measure which includes information on shortest path lengths such as betweenness centrality may be more relevant, even though there is great overlap between hubs found using both metrics on ABIDE data [17]. Future work will take these limitations into consideration and incorporate other methods to determine nodal importance [25, 27].

In this study, we present a novel analysis by combining two complementary frameworks of energy propagation and subnetworks to investigate differences in network efficiency in a large autism and control dataset. Overall, we identify significant group differences in all peripheral subnetworks (feeder, seeder, non-edge) and a preserved central hub in the autism group, further supporting the key role that non-centralised regions play in brain functional organisation. How these energy transport features in the peripheral subnetwork are related to cognitive function and their association with clinical measures in autism remain to be determined.

Acknowledgments. This project has received funding from the American Heart Association and Children's Heart Foundation Postdoctoral Fellowship, 19POST34380005 (AWC) and the European Union's Horizon 2020 research and innovation programme under the Marie Sklodowska-Curie grant agreement No 753896 (MDS).

References

1. Abdelnour, F., Voss, H.U., Raj, A.: Network diffusion accurately models the relationship between structural and functional brain connectivity networks. Neuroimage **90**, 335–347 (2014)
2. Baio, J., et al.: Prevalence of autism spectrum disorder among children aged 8 years-autism and developmental disabilities monitoring network, 11 sites, United States, 2014. MMWR Surveill. Summ. **67**(6), 1 (2018)
3. Chung, A.W., Ahtam, B., Grant, P.E., Im, K.: Rich club-based subnetworks in 16p11.2 deletion syndrome reveal differential structural alterations. In: Organization of Human Brain Mapping, Rome, Italy, p. 5204 (2019)
4. Chung, A.W., Mannix, R., Feldman, H.A., Grant, P.E., Im, K.: Longitudinal structural connectomic and rich-club analysis in adolescent mTBI reveals persistent, distributed brain alterations acutely through to one year post-injury. arXiv:1909.08071 [q-bio.NC], pp. 1–22, September 2019
5. Chung, A.W., Pesce, E., Monti, R.P., Montana, G.: Classifying HCP task-fMRI networks using heat kernels. In: 2016 International Workshop on Pattern Recognition in NeuroImaging (PRNI), pp. 1–4. IEEE (2016)
6. Chung, A.W., et al.: Characterising brain network topologies: a dynamic analysis approach using heat kernels. Neuroimage **141**, 490–501 (2016)
7. Chung, F.R., Graham, F.C.: Spectral Graph Theory, vol. 92. American Mathematical Society (1997)
8. Collin, G., Kahn, R.S., de Reus, M.A., Cahn, W., van den Heuvel, M.P.: Impaired rich club connectivity in unaffected siblings of schizophrenia patients. Schizophr. Bull. **40**(2), 438–448 (2013)
9. Crossley, N.A.: The hubs of the human connectome are generally implicated in the anatomy of brain disorders. Brain **137**(8), 2382–2395 (2014)
10. Di Martino, A., et al.: The autism brain imaging data exchange: towards a large-scale evaluation of the intrinsic brain architecture in autism. Mol. Psychiatry **19**(6), 659 (2014)
11. Gollo, L.L., Zalesky, A., Hutchison, R.M., van den Heuvel, M., Breakspear, M.: Dwelling quietly in the rich club: brain network determinants of slow cortical fluctuations. Philos. Trans. R. Soc. B: Biol. Sci. **370**(1668), 20140165 (2015)

12. Grayson, D.S., et al.: Structural and functional rich club organization of the brain in children and adults. PloS One **9**(2), e88297 (2014)
13. Gu, S., et al.: Controllability of structural brain networks. Nature Commun. **6**, 8414 (2015)
14. van den Heuvel, M.I., et al.: Hubs in the human fetal brain network. Dev. Cogn. Neurosci. **30**, 108–115 (2018)
15. van den Heuvel, M.P., Kahn, R.S., Goñi, J., Sporns, O.: High-cost, high-capacity backbone for global brain communication. Proc. Natl. Acad. Sci. **109**(28), 11372–11377 (2012)
16. Hong, S.J., et al.: Atypical functional connectome hierarchy in Autism. Nature Commun. **10**(1), 1022 (2019)
17. Keown, C.L., Datko, M.C., Chen, C.P., Maximo, J.O., Jahedi, A., Müller, R.A.: Network organization is globally atypical in autism: a graph theory study of intrinsic functional connectivity. Biol. Psychiatry: Cogn. Neurosci. Neuroimaging **2**(1), 66–75 (2017)
18. Ktena, S.I., et al.: Brain connectivity measures improve modeling of functional outcome after acute ischemic stroke. Stroke, published online ahead of print 12, September (2019). https://doi.org/10.1161/STROKEAHA.119.025738
19. Müller, R.A., Fishman, I.: Brain connectivity and neuroimaging of social networks in Autism. Trends Cogn. Sci. **22**, 1103–1116 (2018)
20. Power, J.D., Schlaggar, B.L., Lessov-Schlaggar, C.N., Petersen, S.E.: Evidence for hubs in human functional brain networks. Neuron **79**(4), 798–813 (2013)
21. Raj, A., LoCastro, E., Kuceyeski, A., Tosun, D., Relkin, N., Weiner, M., Initiative ADNI: Network diffusion model of progression predicts longitudinal patterns of atrophy and metabolism in Alzheimer's disease. Cell Rep. **10**(3), 359–369 (2015)
22. Rudie, J.D., et al.: Altered functional and structural brain network organization in autism. NeuroImage: Clin. **2**, 79–94 (2013)
23. Sato, J.R., et al.: Connectome hubs at resting state in children and adolescents: reproducibility and psychopathological correlation. Dev. Cogn. Neurosci. **20**, 2–11 (2016)
24. Schirmer, M.D., Chung, A.W.: Structural subnetwork evolution across the lifespan: rich-club, feeder, seeder. In: Wu, G., Rekik, I., Schirmer, M.D., Chung, A.W., Munsell, B. (eds.) CNI 2018. LNCS, vol. 11083, pp. 136–145. Springer, Cham (2018). https://doi.org/10.1007/978-3-030-00755-3_15
25. Schirmer, M.D., Chung, A.W., Grant, P.E., Rost, N.S.: Network structural dependency in the human connectome across the life span. Netw. Neurosci. 1–15 (2018)
26. Schirmer, M.D., et al.: Rich-Club organization: an important determinant of functional outcome after acute ischemic stroke. Front. Neurol. **10**, 956 (2019). https://doi.org/10.3389/fneur.2019.00956
27. Van Den Heuvel, M.P., Sporns, O.: Rich-club organization of the human connectome. J. Neurosci. **31**(44), 15775–15786 (2011)
28. Van Den Heuvel, M.P., Sporns, O.: A cross-disorder connectome landscape of brain dysconnectivity. Nature reviews. Neuroscience **20**(7), 435–446 (2019). https://doi.org/10.1038/s41583-019-0177-6
29. Varoquaux, G., Craddock, R.C.: Learning and comparing functional connectomes across subjects. NeuroImage **80**, 405–415 (2013)
30. Verhelst, H., Vander Linden, C., De Pauw, T., Vingerhoets, G., Caeyenberghs, K.: Impaired rich club and increased local connectivity in children with traumatic brain injury: local support for the rich? Hum. Brain Mapp. **39**(7), 2800–2811 (2018)
31. Zhang, F., Hancock, E.R.: Graph spectral image smoothing using the heat kernel. Pattern Recogn. **41**(11), 3328–3342 (2008)

A Mass Multivariate Edge-wise Approach for Combining Multiple Connectomes to Improve the Detection of Group Differences

Javid Dadashkarimi[1], Siyuan Gao[2], Erin Yeagle[3], Stephanie Noble[3],
and Dustin Scheinost[3(✉)]

[1] Department of Computer Science, Yale University, New Haven, CT, USA
javid.dadashkarimi@yale.edu
[2] Department of Biomedical Engineering, Yale University, New Haven, CT, USA
[3] Department of Radiology and Biomedical Imaging, Yale School of Medicine,
New Haven, CT, USA

Abstract. Functional connectivity derived from functional magnetic resonance imaging data has been extensively used to characterize individual and group differences. While these connectomes have traditionally been constructed using resting-state data, recent work has highlighted the importance of combining multiple task connectomes, particularly for identifying individual differences. Yet, these methods have not yet been extended to investigate differences at the group level. Here, we propose a mass multivariate edge-wise approach to improve the detection of group differences by combining connectomes from multiple sources. For each edge, the magnitude of connection strength from each of multiple connectomes are included in statistical hypothesis testing. We evaluate the proposed approach by estimating sex differences in two large, publicly available datasets: the Human Connectome Project and Philadelphia Neurodevelopmental Cohort. Results indicate the proposed mass multivariate edge-wise analysis offers improved detection of group differences compared to univariate analysis, and support the utility of combining multiple connectomes to improve detection of group differences.

1 Introduction

Functional connectomics derived from functional magnetic resonance imaging (fMRI) is a powerful framework to elucidate individual and group differences in brain organization [1]. While connectomes are traditionally generated from resting-state data [2], recent work has shown that connectomes generated from task data offer a significant improvement in detecting individual and group differences [3,4]. Furthermore, combining multiple task connectomes per subject further increases the amount of information useful for detecting these differences [5]. However, these methods have only been used in the context of characterizing individual differences, not group differences. Thus, there remains a need to

© Springer Nature Switzerland AG 2019
M. D. Schirmer et al. (Eds.): CNI 2019, LNCS 11848, pp. 64–73, 2019.
https://doi.org/10.1007/978-3-030-32391-2_7

develop methods that combine connectomes from multiple sources in the context of detecting group differences.

To address this need, we extended a traditional mass univariate edge-wise approach (c.f. [6]) to perform multivariate inferences that include, for each edge, the connectivity strengths of all tasks. We label this approach "mass multivariate edge-wise analysis". We compared our multivariate approach for detecting group differences to the traditional mass univariate approach using task connectomes and the general functional connectivity approach for combining multiple task connectomes into a single connectome using the Human Connectome Project (HCP) and Philadelphia Neurodevelopmental Cohort (PNC) datasets [7,8]. We hypothesize that the proposed mass multivariate approach will detect a greater number of edges exhibiting significant sex effects in comparison to the competing methods. Together, our results support the utility of combining multiple task connectomes to improve the detection of group differences.

2   Related Works

Efforts aimed at combining connectomes from multiple sources is an active field of research. Although an exhaustive review of this field is outside the scope of this paper, we briefly highlight some relevant work. Several studies have combined structural connectomes from diffusion tensor imaging (DTI) with functional connectomes [9,10]. Similarly, work has been done to combine connectomes derived from electroencephalogram (EEG) and fMRI data [11]. Yet, combining multiple connectomes from different tasks has received less attention. To our knowledge only three approaches, based on canonical correlation analysis, ridge regression, and averaging connectomes, respectively, have been proposed [3,5].

3   Methods

In this section, we derive our proposed mass multivariate edge-wise analysis. First, we briefly review multivariate hypothesis testing using the Hotelling's T^2 test in Sect. 3.1; second, we discuss univariate edge-wise analysis in Sect. 3.2; and, finally, we propose our mass multivariate edge-wise analysis in Sect. 3.3. Figure 1 shows an overall schematic of the two analyses.

3.1   Hotelling's T^2 Test

Briefly, the t-test is a statistical method to determine if the means of data from two different groups differ from each other. The Hotelling's T^2 test is a generalization of the t-test that allows for multivariate, rather than univariate, hypothesis testing. For a t-distribution, a confidence interval for a sample of size n, standard deviation of s, and significance level of α is defined as:

$$\bar{x} \pm t_{1-\alpha/2,n-1}\frac{s}{\sqrt{n}} \tag{1}$$

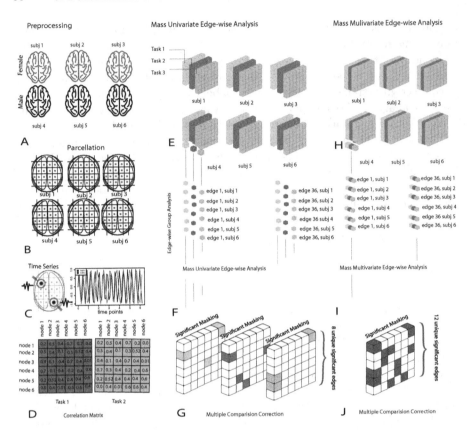

Fig. 1. Mass multivariate edge-wise analysis pipeline. *Common preprocessing steps:* **(A)** fMRI data consisting of k tasks acquired from two groups (*e.g.* males and female) of participants; **(B)** parcellate the brain into N nodes; **(C)** average timeseries for each node; **(D)** generate connectomes $\mathbf{x}_i \in \mathbb{R}^{N \times N}$ for each task and participant using the time series; *Mass Univariate Edge-wise Analysis:* **(E)** independently for each task connectome, create a vector connectivity strength across all participants for each edge; **(F)** apply a t-test on these vectors and corresponding labels $y \in \{0, 1\}$; **(G)** perform hypothesis testing and multiple comparison correction using *a priori* thresholds; *Mass Multivariate Edge-wise Analysis:* **(H)** stack all k connectomes $\mathbf{X} \in \mathbb{R}^{k \times N \times N}$, creating a matrix of connectivity strength across all participants for each; **(I)** apply a Hotelling's T^2 on $\mathbf{X}[:, i, j]$ and $y \in \{0, 1\}$; **(J)** perform hypothesis testing and multiple comparison correction using *a priori* thresholds.

where $t_{1-\alpha/2, n-1}$ is $1 - \alpha/2$ fraction of t-distribution with $n - 1$ degrees of freedom. In other words, with $1 - 2 \times \frac{\alpha}{2}$ trials, the true mean spans in this interval. To test if the sample mean has a value of μ_0 based on null-hypothesis, we need to verify $T = (\bar{x} - \mu_0)/(s/\sqrt{n}) < t_{\alpha/2, n-1}$ or $T = (\bar{x} - \mu_0)/(s/\sqrt{n}) > t_{1-\alpha/2, n-1}$. We can re-write $T^2 = n \frac{(\bar{x} - \mu_0)^2}{s^2}$ and compare with squared form of t-distribution.

Given that $\bar{\mathbf{x}} = (\bar{x}_1, \bar{x}_2, .., \bar{x}_k)$ is a vector of k normally distributed variables and $\boldsymbol{\mu}_0 = (\mu_1^0, \mu_2^0, .., \mu_k^0)$ are the corresponding means, we define:

$$T^2 = n(\bar{\mathbf{x}} - \boldsymbol{\mu}_0)\mathbf{S}^{-1}(\bar{\mathbf{x}} - \boldsymbol{\mu}_0) \tag{2}$$

where $\mathbf{S}$ is the variance-covariance matrix. Equation 2 is exactly comparable to the ratio of between group variance $n(\bar{\mathbf{x}} - \boldsymbol{\mu}_0)(\bar{\mathbf{x}} - \boldsymbol{\mu}_0)/(m-1)$ and within group variance $\sum_i \mathbf{S}_i/m$. For hypothesis testing, Hotelling's T^2 is first transformed into an F-statistic using $F = \frac{n_1+n_2-p-1}{p(n_1+n_2-2)}T^2 \sim F_{p,n_1+n_2-p-1}$. The null hypothesis at a chosen significance level is rejected if the calculated value is greater than the F-table critical value. Rejecting the null hypothesis means that at least one of the parameters, or a combination of one or more parameters working together, is significantly different between the groups.

While mathematically similar, a Hotelling's T^2 test has a major advantage over a t-test [12]. Since a single comparison is made in the former test, the Type I error rate is well controlled and the relationship between multiple variables is taken into account. In summary, a t-test will denote which variables differ between groups; while a Hotelling's T^2 summarizes the between-group differences.

3.2   Mass Univariate Edge-wise Analysis

Using a single connectome for each participant as input, mass univariate edge-wise analysis involves performing a statistical test (typically, a t-test) to independently compare groups for each edge in a single connectome [6]. This results in a "difference matrix" of test statistics, representing the magnitude of group difference at each edge of that connectome. Multiple comparison correction for the $\frac{N \times (N-1)}{2}$ comparisons needs to be applied, where N is the number of nodes in the parcellation. Finally, for each edge, the null hypothesis can be rejected if the test statistic is greater than the critical value. Figure 1E-G shows an overview of the mass univariate edge-wise analysis.

3.3   Mass Multivariate Edge-wise Analysis

The proposed mass multivariate edge-wise analysis is a multivariate extension of the univariate approach. Using multiple task connectomes for each participant as inputs, this multivariate analysis involves performing a multivariate test (*e.g.* Hotelling's T^2) on connectivity strength for all tasks to create a "difference matrix" of test statistics. In this manner, the connectivity strength of an edge for each task is included in statistical testing. The "difference matrix" can then be corrected for multiple comparisons as above and thresholded for statistical significance. Post-hoc univariate tests (*e.g.* t-test) can be performed on each individual connectome to determine which task or tasks most likely contributed to edges exhibiting significant differences. Figure 1H–J shows an overview of the proposed mass multivariate edge-wise analysis.

4   Experiments

4.1   Datasets

We used two standard datasets in our analysis: the Human Connectome Project (HCP) and Philadelphia Neurodevelopmental Cohort (PNC) [7,8] (see Table 1). We narrowed the participants into the set of participants with mean frame-to-frame displacement less than 0.1 mm and maximum frame-to-frame displacement of less than 0.15 mm. The HCP dataset consists of 9 tasks: gambling (gam), emotion, language, motor, relation, social, working memory (wm), and two resting-state runs (rest1, rest2). The PNC dataset consists of three tasks: emotion, wm, and a resting-state run.

Table 1. Characteristics for the HCP and PNC datasets.

ID	Collection	#male	#female	Size	age	#tasks
HCP	Human Connectome Project	241	274	515	28 ± 3.98	9
PNC	Philadelphia Neurodevelopmental Cohort	251	320	571	15 ± 3.65	3

4.2   Preprocessing

For the HCP dataset, we started with the minimally preprocessed HCP data [13]. For the PNC dataset, functional images were slice-timed and motion-corrected and registered into common space as previously described [4]. Further preprocessing steps were performed using BioImage Suite [14]. These included regressing 24 motion parameters, regressing the mean time courses of the white matter, CSF, and grey matter, removing the linear trend, and low-pass filtering.

Regions were delineated according to the Shen atlas [15]. This atlas, defined in an independent dataset, provides a parcellation of the whole gray matter (including subcortex) into 268 contiguous, functionally coherent regions. These nodes have also been grouped into 10 functionally coherent "networks". For each scan, the average timecourse within each region was obtained, and the Pearson's correlation between the mean timecourses of each pair of regions was calculated. These correlation values provided the edge strengths for a 268×268 symmetric correlation matrix for each combination of subject, session, and run. These correlations were converted to be approximately normally distributed using a Fisher transformation.

4.3   Evaluation and Competing Methods

Using the HCP and PNC datasets, we evaluated our mass multivariate edge-wise analysis by quantifying the number of edges exhibiting significant differences between males and females. Significance was defined as $q < 0.005$ using the Storey procedure for positive False Discovery Rate (pFDR) correction [16].

We compared the number of significant edges between male and female participants detected by our mass multivariate edge-wise analysis to the number of significant edges detected by three competing approaches. First, we performed a standard mass univariate edge-wise analysis (as described in Sect. 3.2) for each task connectome, independently. For each task, we quantified the number of edges that exhibited significant differences after correcting for multiple comparisons using pFDR. These results provide a baseline for the amount of significant differences detectable in each connectome. Second, we quantified the union of the significant edges from each task from the first comparison approach. This result produces a comparison for the number of significant edges when naively combining information across all tasks. Third, we combined all task connectomes using the general functional connectivity method from [3] (*i.e.* averaging all connectomes), performed a standard mass univariate edge-wise analysis on this averaged connectome, and quantified the number of edges that exhibited significant differences after correcting for multiple comparisons using pFDR. This result produced a comparison for the number of significant edges when combining information across all tasks with a previously published method [3].

We would like to note that results based on independently performed mass univariate edge-wise analyses (*i.e.* the first and second competing approaches from above) do properly control for type I error as multiple comparison correction is only performed on independent analyses, not accounting the multiple tasks. This will inflate the number of edges detected with these approaches. However, type I error is well controlled for our mass univariate edge-wise analysis as connectomes are combined for a single comparison.

4.4   Visualization of Anatomical Locations of Significant Edges

To visualize anatomical locations of significant edges, we used stacked area plots to explain the probability of finding a significant edge in the network of interest. The 268 nodes were grouped into 10 networks for visualization: limbic system (Limb), default mode network (DMN), cerebellum (CBL), basal ganglia (BG), mediofrontal cortex (MF), motor areas (Mot), subcortical areas (Sc), visual association (VAs), visual-I (VI), visual-II (VII). Within a network of M nodes, the hypergeometric distribution gives the probability of finding a sex difference among all $\binom{M}{2}$ possible edges [17,18]). This is exactly analogous to the traditional example of drawing, without replacement, a white ball from a bag of $\binom{M}{2}$ white balls and $\binom{N}{2} - \binom{M}{2}$ black balls, where M is the number of nodes in a given network and N is the total number of nodes in the connectome.

5   Results

First, we compared the number of edges that were found to significantly differ between male and female participants as a result of the multivariate analysis with the number of edges that differed for each individual task in the univariate analysis (Table 2). A greater number of edges were found to differ between groups when combining all connectomes compared to any single connectome. In the HCP dataset, the proposed multivariate approach detected significant sex effects in approximately 28% of the total number of edges; whereas, in the PNC dataset, the proposed multivariate approach detected significant sex effects in approximately 4.5% of the total number of edges.

Next, we compared the number of significant edges for the proposed multivariate approach with two competing approaches that also combine all connectomes (Table 3). Our mass multivariate approach resulted in the greatest number of significant edges in HCP dataset. Yet a mass univariate edge-wise analysis on the mean connectome across tasks resulted in the greatest number of significant edges in the PNC dataset.

Table 2. The number of edges exhibiting significant differences between male and female participants for both univariate and multivariate approaches. For each task in the univariate approach, differences between groups were calculated for each connectome individually. For the multivariate approach, differences between groups were calculated using all task connectomes together. Significance was defined as $q < 0.005$ corrected for multiple comparisons using pFDR. Bold numbers show the rows with greatest number of significant edges.

	Task	#sig	%conn		Task	#sig	%conn		Task	#sig	%conn
HCP	gam	1355	3.79%	HCP	motor	1095	3.06%	PNC	emotion	1131	3.16%
	rest1	3331	9.31%		relation	716	2.00%		wm	883	2.47%
	rest2	2207	6.17%		social	1069	2.99%		rest	225	0.63%
	language	1315	3.67%		wm	1873	5.23%		multivariate	**1592**	**4.45%**
	emotion	1695	4.74%		multivariate	**9940**	**27.78%**				

Table 3. The number of significant edges for the 'union' (univariate) column are based on the union of all significant edges each individual univariate edge-wise analysis. The number of significant edges for the 'mean' (also univariate) column are based on the univariate edge-wise analysis using the average connectome of all tasks. The number of significant edges for the multivariate column are based on the proposed mass multivariate edge-wise analysis. Significance was defined as $q < 0.005$ corrected for multiple comparisons using pFDR. Bold numbers show the approaches with greatest number of significant edges.

		Union	Mean	Multivariate
HCP	#sig	8091	5701	**9940**
	%conn	22.61%	15.93%	**27.78%**
PNC	#sig	1740	**1920**	1590
	%conn	4.86%	**5.36%**	4.44%

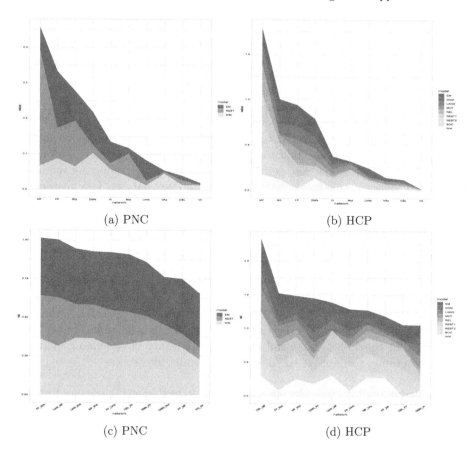

(a) PNC   (b) HCP

(c) PNC   (d) HCP

Fig. 2. Area plots for within networks Fig. 2a and b, and between networks Fig. 2c and d. The horizontal axis shows network names and vertical axis shows the value of the hypergeometric function [17, 18]. Networks are sorted according to the cumulative values across tasks.

Last, we investigated the anatomical locations of the significant edges detected using the proposed approach. The MF, FP, and BG networks had the greatest total number of significant edges, while the VAs and VII networks had the fewest (Fig. 2a–b). The DMN and FP showed the most consistent between-network association in the HCP dataset, while the Limb and MF networks show the most consistent between network associations in PNC dataset (Fig. 2c–d).

6   Discussion and Conclusions

In this work, we proposed a method to combine connectomes from multiple sources in the context of detecting group differences. To accomplish this, we extended a traditional mass univariate edge-wise analysis by incorporating mul-tivariate statistics, where, for each edge, the connectivity strength for each task is included in hypothesis testing. While connectomes derived from fMRI are typically generated from resting-state data, in this paper, we have shown that combining connectomes generated from multiple sources increases the amount of information useful for characterizing group differences. These results are in agreement with the recent work which has shown that connectomes generated from task data offer a significant improvement in detecting individual and group differences [3,4]. Furthermore, our results suggests that combining multiple connectomes derived from tasks that tap into multiple cognitive dimensions offers greater power than using a connectome from a single source. Although our mass multivariate edge-wise analysis performed better than the competing methods in the HCP dataset, this was not the case for the PNC dataset. One reason for these diverging results is the higher number of tasks in the HCP compared to the PNC dataset. Future work will include investigations into the optimal number of connectomes for mass multivariate edge-wise analysis. Additional future work will involve extending our mass multivariate edge-wise analysis to use multivariate analysis of variance (MANOVA)–the multivariate extension of an analysis of variance (ANOVA)–to compare multiple groups. Finally, our mass multivariate edge-wise analysis generalizes across different sources of connectomes as long as all connectomes have the same size. We will explore incorporating structural connectomes generated from DTI data and functional connectomes generated from EEG data. In conclusion, our results support the utility of combining multiple task connectomes to improve the detection of group differences.

Acknowledgements. Data were provided in part by the Human Connectome Project, WU-Minn Consortium (Principal Investigators: David Van Essen and Kamil Ugurbil; 1U54 MH091657) funded by the 16 NIH Institutes and Centers that support the NIH Blueprint for Neuroscience Research; and by the McDonnell Center for Systems Neuroscience at Washington University. The remainder of the data used in this study were provided by the Philadelphia Neurodevelopmental Cohort (Principal Investigators: Hakon Hakonarson and Raquel Gur; phs000607.v1.p1). Support for the collection of the data sets was provided by grant RC2MH089983 awarded to Raquel Gur and RC2MH089924 awarded to Hakon Hakonarson. All subjects were recruited through the Center for Applied Genomics at The Children's Hospital in Philadelphia.

References

1. Dubois, J., Adolphs, R.: Building a science of individual differences from fMRI. Trends Cogn. Sci. **20**(6), 425–443 (2016)
2. Song, X.-W., et al.: REST: a toolkit for resting-state functional magnetic resonance imaging data processing. PLoS ONE **6**(9), e25031 (2011)
3. Elliott, M.L.: General functional connectivity: shared features of resting-state and task fMRI drive reliable and heritable individual differences in functional brain networks. NeuroImage **189**, 516–532 (2019)
4. Greene, A.S., Gao, S., Scheinost, D., Constable, R.T.: Task-induced brain state manipulation improves prediction of individual traits. Nat. Commun. **9**(1), 2807 (2018)

5. Gao, S., Greene, A., Constable, T., Scheinost, D.: Combining multiple connectomes improves predictive modeling of phenotypic measures. Neuroimage **201**, 116038 (2019)
6. Zalesky, A., Fornito, A., Bullmore, E.T.: Network-based statistic: identifying differences in brain networks. NeuroImage **53**(4), 1197–1207 (2010)
7. Van Essen, D.C., et al.: The WU-Minn human connectome project: an overview. Neuroimage **80**, 62–79 (2013)
8. Satterthwaite, T.D., et al.: The Philadelphia neurodevelopmental cohort: a publicly available resource for the study of normal and abnormal brain development in youth. Neuroimage **124**, 1115–1119 (2016)
9. Ystad, M.: Cortico-striatal connectivity and cognition in normal aging: a combined DTI and resting state fMRI study. Neuroimage **55**(1), 24–31 (2011)
10. Wang, J., et al.: Alternations in brain network topology and structural-functional connectome coupling relate to cognitive impairment. Front. Aging Neurosci. **10**, 404 (2018)
11. Deligianni, F., Centeno, M., Carmichael, D.W., Clayden, J.D.: Relating resting-state fMRI and EEG whole-brain connectomes across frequency bands. Front. Neurosci. **8**, 258 (2014)
12. Hotelling, H., et al.: A generalized t test and measure of multivariate dispersion. In: Proceedings of the Second Berkeley Symposium on Mathematical Statistics and Probability. The Regents of the University of California (1951)
13. Glasser, M.F., et al.: The minimal preprocessing pipelines for the human connectome project. Neuroimage **80**, 105–124 (2013)
14. Joshi, A.: Unified framework for development, deployment and robust testing of neuroimaging algorithms. Neuroinformatics **9**(1), 69–84 (2011)
15. Shen, X., Tokoglu, F., Papademetris, X., Constable, R.T.: Groupwise whole-brain parcellation from resting-state fMRI data for network node identification. Neuroimage **82**, 403–415 (2013)
16. Storey, J.D.: A direct approach to false discovery rates. J. R. Stat. Soc.: Ser. B (Stat. Methodol.) **64**(3), 479–498 (2002)
17. Abramowitz, M., Stegun, I.A.: Handbook of Mathematical Functions: With Formulas, Graphs, and Mathematical Tables, vol. 55. Courier Corporation, Chelmsford (1965)
18. Luke, Y.L.: Special Functions and their Approximations, vol. 2. Academic press, Cambridge (1969)

Adversarial Connectome Embedding for Mild Cognitive Impairment Identification Using Cortical Morphological Networks

Alin Banka and Islem Rekik$^{(\boxtimes)}$

BASIRA Lab, Faculty of Computer and Informatics,
Istanbul Technical University, Istanbul, Turkey
irekik@itu.edu.tr

Abstract. *Cortical Morphological Networks* provided unprecedented insights into connectional brain alternations in patients diagnosed with *mild cognitive impairment* (MCI) and, in combination with deep learning techniques, they can further be utilized to build computer-aided MCI diagnosis models. In this paper, we introduce *Adversarial Connectome Embedding* (ACE) architecture, which is rooted in graph convolution and adversarial regularization to learn relevant connectional features for MCI classification. Existing connectome-based embedding methods for examining the healthy and disorder brain connectivity generally rely on vectorizing the connectivity matrix and use typical Euclidean embedding methods (e.g., principal component analysis), which work best in Euclidean spaces such as images. On the other hand, a connectome, which is modeled as a brain graph or network, lies in a non-Euclidean space. Hence, the connectome vectorization might cause losing its topological structure which can be leveraged to boost brain graph classification for diagnosis. To fill this gap, we leverage *geometric deep learning*, a nascent field which extends deep Euclidean feature representation learning to non-Euclidean spaces. First, we propose to use a *geometric* autoencoder with graph convolutional layers to learn a latent brain connectivity representation (i.e., embedding) that exploits the connectome topology. Secondly, we utilize an adversarial regularizing network which forces the learned latent distribution to match the prior distribution of the connectomes. Finally, we feed the adversarially regularized latent connectome embeddings to train a linear classifier for diagnosing MCI patients. ACE achieved the best classification results across different connectomic datasets for MCI versus Alzheimer's disease classification in comparison with typical graph embedding techniques.

Keywords: Adversarial graph embedding · Cortical morphological networks · Mild cognitive impairment diagnosis · Geometric deep learning · Brain connectivity classification

© Springer Nature Switzerland AG 2019
M. D. Schirmer et al. (Eds.): CNI 2019, LNCS 11848, pp. 74–82, 2019.
https://doi.org/10.1007/978-3-030-32391-2_8

1   Introduction

Dementia is an umbrella term used to describe a decline in mental ability, which may result in loss of capacity to complete daily tasks. The most common type of dementia is Alzheimer's Disease (AD), accounting for 60 to 80% of cases. The evolution of the disease is divided into three stages, with the late stage being the most severe. In the final stage of the disorder individuals may lose awareness of their surroundings, experience difficulties in physical activities such as walking, sitting and swallowing, have problems in communicating properly, and are, therefore, in need of constant personal care [1]. Due to the severity of these symptoms, it is a crucial task that early diagnosis of dementia, especially late mild cognitive impairment (LMCI), is given to the patient in order to prevent progression into AD. In order to achieve this, machine learning methods have utilized magnetic resonance imaging (MRI) scans, providing an effective and non-invasive facility to diagnose a variety of neurological disorders. In particular, the connectome is a graph-based representation of the brain wiring derived from MRI data such as resting-state functional MRI (rsfMRI). Furthermore, the connectome allows to efficiently map neural connections within an organism's nervous system and provides an effective and systematic way to extract *connectional* features encoding the relationship between pairs of regions of interest (ROIs) of the brain in both health and disease. These features can be utilized in numerous applications regarding medical imaging and can provide an enormous assistance to medical professionals working in diagnostics of neurological disorders to develop connectivity-targeting treatments.

Brain disorders alter the brain construct in neural activity, which can be quantified using functional magnetic resonance imaging (MRI), as well as brain morphology, which can be measured using structural T1-weighted MRI. Therefore, constructing a model that can accurately distinguish MCI from AD will enable the development of precise treatments for each form of brain dementia. Nevertheless, disentangling dementia brain states remains a challenging classification problem. To mitigate this issue, examining cortical measures derived from the multi-folded surface of the cerebral cortex for dementia state identification, such as the cortical thickness, has been presented in several previous studies [2,3]. For example, Frisoni *et al.* demonstrated that AD subjects show reduction in cortical thickness compared with control subjects [4]. Nevertheless, these methods were reliant on volumetric thickness analysis, despite there being that AD alters the shape of cortical regions as well [5]. As a result, other studies explored cortical thickness using surface-based methods involving spectral shape description [6], or combining shape-derived features with voxel features [7]. However, these techniques only considered vertex-level morphological features. To address these limitations, recent works [8–11] proposed cortical morphological networks (CMN) derived from T1-w MRI and estimated using different cortical attributes (e.g., cortical thickness) to model the co-regional changes in brain morphology. Although CMN presents a higher order representation of brain morphological changes in comparison with region-wise cortical measurements, it remains a *shallow* graph-based representation of the brain complexity as a system.

Most of the existing studies overlooked the *deep and hierarchical* topological properties of brain connectivity. In fact, traditional classification methods use feature extraction from connectome vectorization and then train a classifier (e.g., support vector machines (SVM)) for classification. These techniques rely on mathematical transformations that do not take advantage of the topological structure of the graph. Recently, new methods, especially in the emerging field of *geometric deep learning*, such as *Graph Convolutional Networks* [12], attempted to solve this issue by introducing a new type of convolutional and deconvolutional layers, which operate *directly* on graphs, thus learning deep graph embeddings that are more in line with the graph representation. In particular, more recently, adversarially regularized graph autoencoder for graph embedding [13] was introduced, which designs an autoencoder using graph convolutional layers to extract latent codes (i.e., embeddings) combined with an adversarial regularization to exploit the features. Inspired from [13], we propose Adversarial Connectome Embedding (ACE), which learns a deep connectome representation via a graph-based encoding and decoding steps. Specifically, ACE comprises stacked graph convolutional layers and a discriminator network, normally included in adversarial methods, in order to force the distribution of the learned latent embeddings to match the distribution of the input data. Next, we leverage the learned connectome embeddings to train a linear SVM classifier using different CMNs. Our method boosted LMCI/AD classification accuracy compared to traditional methods such as connectome vectorization and using a Euclidean dimensionality reduction technique (e.g., principal component analysis (PCA)).

2   Method

In this section, we elaborate over the details of our proposed method. In Fig. 1, we display a chart of ACE classification pipeline. Our method is based on three fundamental steps: (**A**) learning of latent connectome embedding, which consists of an autoencoder with stacked graph convolutional layers, (**B**) adversarial regularization of the distribution of the latent embeddings, which is achieved by using a discriminator network, commonly used in generative adversarial networks (GAN) Networks (GANs), and (**C**) classification using linear SVM, which is a commonly-known method for classification.

A- Connectome Embedding Learning. In this step, we root our framework in the recently designed graph-based autoencoder network [13], which is an unsupervised graph representation learning technique. The proposed ACE network (Fig. 1) takes brain connectivity matrices as inputs and learns their corresponding embeddings via a graph convolution-based encoder, which are next decoded using a graph deconvolution based decoder for brain connectivity reconstruction. Specifically, the encoding network operates as a Graph Convolutional Network (GCN) and is created by stacking Graph Convolutional Layers. These Layers introduce learnable filters utilized for feature extraction on graphs and they are

Brain Connectome Classification Using Adversarially Embedded Connectomes (ACE)

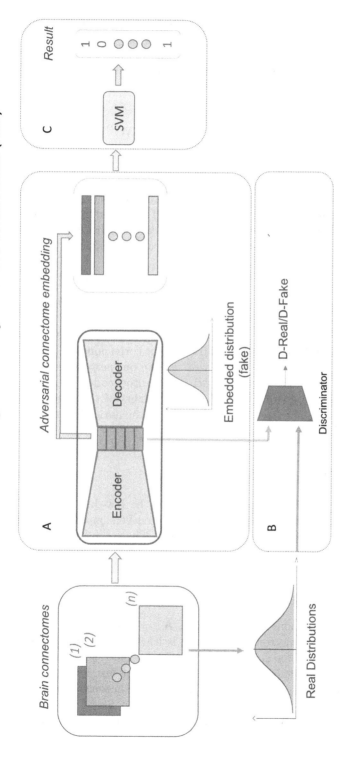

Fig. 1. *Adversarial Connectome Embedding (ACE) Learning.* **(A) Connectome embedding learning:** We sequentially train the autoencoder over 5 epochs for each sample and learn the deep embedded representation of the input connectome (i.e., the latency codes). **(B) Adversarial Regularization.** The learned embeddings are regularized through the discriminator D. A prior distribution is sampled from the input connectomes and is utilized to force the learned latent distribution to match the input distribution. The discriminator D is defined as a multilayer perceptron. **(C) Classification using SVM.** The learned and regularized connectome embeddings are then used as input feature vectors to a linear support vector machine (SVM), which is trained to learn to distinguish between two connectomic groups (e.g., late mild cognitive impairment versus Alzheimer's disease).

constructed to adapt in a way that considers the topological construction of a graph, rather than disregarding it completely such as filters used for Euclidean data would normally do. GCN functions are defined using convolutions of graph data in the spectral domain to learn a layer-wise representation expressed by a spectral convolution function $f_\phi(\mathbf{Z}^{(l)}, \mathbf{A} | \mathbf{W}^{(l)})$.

The function f_ϕ is specified as follows:

$$f_\phi(\mathbf{Z}^{(l)}, \mathbf{A} | \mathbf{W}^{(l)}) = \phi(\widetilde{\mathbf{D}}^{-\frac{1}{2}} \widetilde{\mathbf{A}} \widetilde{\mathbf{D}}^{-\frac{1}{2}} \mathbf{Z}^{(l)} \mathbf{W}^{(l)})$$

Where ϕ is the activation function of the $(l)^{th}$ layer. $\widetilde{\mathbf{A}} = \mathbf{A} + \mathbf{I}$, where $\mathbf{I}$ is the identity of the adjacency matrix $\mathbf{A}$, and $\widetilde{\mathbf{D}}_{ii} = \sum_j \widetilde{\mathbf{A}}_{ij}$. Concerning our problem, we are more interested in learning topological features, since the elements of the input connectivity matrices represent morphological dissimilarity between brain regions of interest (ROIs) and our nodes contain no features. Graph convolutional layers are suitable for this application. Specifically, in our encoding part of the network we specify two layers as follows:

$$\mathbf{Z}^{(1)} = f_{ReLU}(\mathbf{Z}, \mathbf{A} | \mathbf{W}^{(0)}); \qquad \mathbf{Z}^{(2)} = f_{linear}(\mathbf{Z}^{(1)}, \mathbf{A} | \mathbf{W}^{(1)}),$$

Where $\mathbf{W}$ is the weight matrix used to learn the graph convolution filter for each layer. $\mathbf{Z}^{(1)}$ and $\mathbf{Z}^{(2)}$ are the learned embeddings of the corresponding layers $l = 1$ and $l = 2$. In addition, the activation function used in our network is a rectified linear unit (ReLU (.)). On the decoding part of the autoencoder, we attempt to reconstruct the image by computing each of the weights of the edges between nodes i and j of the connectivity matrix, which is mathematically formalized as follows:

$$Dec(\hat{\mathbf{A}} | \mathbf{Z}) = sigmoid(\mathbf{z}_i^\top, \mathbf{z}_j)$$

Where $\mathbf{z}$ is the learned connectome embedding in the low-dimensional space, $\mathbf{z}^T$ denotes its transpose and $\hat{\mathbf{A}}$ is the reconstructed version of $\mathbf{A}$. We apply sigmoid function to obtain the reconstructed graph. The reconstruction error to minimize is calculated by:

$$\mathbf{L}_0 = \mathbf{E}_{q(\mathbf{Z} | \mathbf{X}, \mathbf{A})}[\log P(\widetilde{\mathbf{A}} | \mathbf{Z})]$$

Where the graph convolutional encoder $q(\mathbf{Z} | \mathbf{X}, \mathbf{A}) = G(\mathbf{Z}, \mathbf{A})$ encodes the representation $\mathbf{Z}^{(2)}$. Each connectome embedding is learned in such a way so that the reconstructed connectome is as close as possible to the original connectome. Nevertheless, this form of training can lead to the latent distribution deviating immensely from the prior distribution. The usage of the autoencoder alone may lead to overfitting and thus lower accuracy rates for our model. Consequently, a form of regularization is required to mitigate the overfitting and boost the accuracy rates.

B- Adversarial Regularization. In the spirit of [13], we force the latent embeddings to match the prior distribution by using adversarial regularization. We achieve this by using a discriminator network, which is based on the multilayer perceptron (MLP) and is constructed from regular dense layers, where the final layer consists of only one output. A discriminator is often employed in GANs where it is trained to discriminate between the created samples provided by the generator and real samples of the dataset, thus, forcing the constructed samples to resemble the real ones. In our case, we are trying to force the latent distribution to resemble the prior distribution. Therefore, the output of the discriminator is used to distinguish whether a distribution comes from a latent embeddings (fake) or a prior distribution (real). This *regularizes* the learned embedding and *aligns* it better with the data prior distribution, which results in lower reconstruction errors for the test set. This regularization step is based on a general cross-entropy loss function, which is typically adopted in adversarial models. The loss is computed as follows:

$$-\frac{1}{2}\mathbf{E}_{z\sim p_z}[\log D(\mathbf{Z})] - \frac{1}{2}\mathbf{E}_X[\log(1 - D(G(\mathbf{X}, \mathbf{A})))]$$

By combining the autoencoder with the discriminator, the training process can be defined as:

$$\min_G \max_D \mathbf{E}_{z\sim p_z}[\log D(\mathbf{Z})] + \mathbf{E}_{z\sim p_z}[\log(1 - D(G(\mathbf{X}, \mathbf{A})))]$$

Where $G(\mathbf{X}, \mathbf{A})$ denotes the generator, $\mathbf{X}$ is the node feature matrix, and $D(\mathbf{Z})$ is the discriminator. The combined losses produce a more refined and optimized graph embeddings, providing a deep and more relevant representation of the brain connectome.

C- Classification using SVM. Next, we train a linear SVM classifier to classify the learned embeddings into two neurological classes: LMCI and AD, which are difficult to disentangle. In the testing step, we use the trained SVM model to predict the label for an input connectome embedding.

3   Results and Discussion

Dataset and Benchmarking: In our study, we evaluated our model over 77 subjects (41 AD and 36 LMCI) from ADNI GO public dataset, each with structural T1-w MR image [14]. Each subject has 3 cortical morphological networks, each encoded in a symmetric adjacency matrix with size (35×35) [8] and derived using a specific cortical attribute: (1) maximum principal curvature, (2) cortical thickness, and (3) sulcal depth. We trained our classifier with only CMN derived from the left hemisphere. ACE is trained sequentially over 5 epochs for each graph. The model is benchmarked against two other methods which extract features by vectorizing the brain connectivity matrices. One of the model uses plain SVM, whereas the other one leverages Euclidean feature reduction method prior to SVM training by utilizing principal component analysis (PCA). In our results, we report the best PCA+SVM performance by varying the number of dimensions.

Training and Testing Randomization Strategy. We arbitrarily divide the dataset into training and testing sets, where the testing set is roughly one fifth of the total dataset. The models are trained and tested by continuously shuffling and splitting of the dataset over 100 runs. In Fig. 2, the first three charts display the average classification accuracy of each method over the 100 runs for each of the 3 CMNs and in the final bar plots we report the average accuracy across all CMNs.

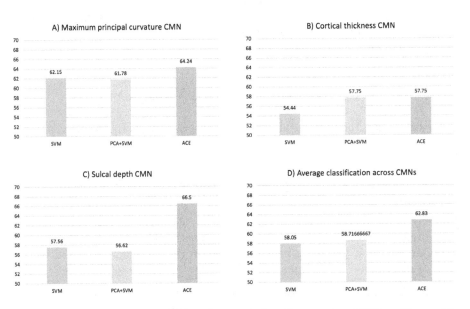

Fig. 2. *Comparison between ACE and vectorized connectomes + SVM –with or without PCA.* We display the average classification accuracy for each cortical morphological network (CMN) and the average for the three models. ACE achieves the best classification results in comparison with SVM and PCA+SVM.

Evaluation and Comparison: We compare our method with two other models: (A) vectorized connectome + SVM (B) vectorized connectome + PCA + SVM. For connectome vectorization, we simply extract the connectivity values in the off-diagonal upper triangular part of each symmetric connectivity matrix. Next, we directly use the vectors for SVM training and classification or we perform PCA dimensionality reduction in order to extract relevant features. We use PCA for dimensionality reduction after we tested it for different numbers of components and compare our new method with only the highest performing version of PCA + SVM. The latent embeddings from ACE are learned using sequential learning. Then, we feed the latent embeddings as inputs for SVM training, which predicts the unknown class for the testing set. Clearly, ACE outperforms the other two methods in two CMNs and matches PCA+SVM in one CMN. On average, ACE achieves the best classification accuracy in comparison with baseline methods. This paper is a first proof-of-concept connectome graph embedding leveraging geometric deep learning outperforms typical Euclidean data embeddings techniques such as PCA for improving brain state classification results.

4   Conclusion

In this paper, we proposed an adversarial connectome embedding framework rooted in geometric deep learning for late dementia states classification using cortical morphological networks. Our framework includes an autoencoder, which consists of graph convolutional layers and adversarial embedded distribution regularization. The usage of graph convolutional layers for deep graph embedding allows the learning of deep representative features of the connectome without discarding its topological properties. Our model outperformed the classical models which use connectome vectorization, resulting in a loss of the topology of the connectome. Building on this work and inspired by graph convolution and adversarial learning, we will further extend the designed architecture to handle multi-view brain networks for joint embedding and classification.

References

1. Gaugler, J., James, B., Johnson, T., Marin, A., Weuve, J.: Alzheimer's disease facts and figures. Alzheimers Dement. **15**(2019), 321–387 (2019)
2. McEvoy, L.K., et al.: Alzheimer disease: quantitative structural neuroimaging for detection and prediction of clinical and structural changes in mild cognitive impairment. Radiology **251**, 195–205 (2009)
3. Ridgway, G.R., et al.: Early-onset Alzheimer disease clinical variants: multivariate analyses of cortical thickness. Neurology **79**, 80–84 (2012)
4. Frisoni, G., et al.: Detection of grey matter loss in mild Alzheimer's disease with voxel based morphometry. J. Neurol. Neurosurg. Psychiatry **73**, 657–664 (2002)
5. Kälin, A.M., et al.: Subcortical shape changes, hippocampal atrophy and cortical thinning in future Alzheimer's disease patients. Front. Aging Neurosci. **9**, 38 (2017)
6. Shakeri, M., Lombaert, H., Tripathi, S., Kadoury, S.: Deep spectral-based shape features for Alzheimer's disease classification. In: Reuter, M., Wachinger, C., Lombaert, H. (eds.) SeSAMI 2016. LNCS, vol. 10126, pp. 15–24. Springer, Cham (2016). https://doi.org/10.1007/978-3-319-51237-2_2
7. Tripathi, S., Nozadi, S.H., Shakeri, M., Kadoury, S.: Sub-cortical shape morphology and voxel-based features for Alzheimer's disease classification. In: 2017 IEEE 14th International Symposium on Biomedical Imaging (ISBI 2017), pp. 991–994 (2017)
8. Mahjoub, I., Mahjoub, M.A., Rekik, I.: Brain multiplexes reveal morphological connectional biomarkers fingerprinting late brain dementia states. Sci. Rep. **8**, 4103 (2018)

9. Lisowska, A., Rekik, I.: Pairing-based ensemble classifier learning using convolutional brain multiplexes and multi-view brain networks for early dementia diagnosis. In: Wu, G., Laurienti, P., Bonilha, L., Munsell, B.C. (eds.) CNI 2017. LNCS, vol. 10511, pp. 42–50. Springer, Cham (2017). https://doi.org/10.1007/978-3-319-67159-8_6

10. Soussia, M., Rekik, I.: Unsupervised manifold learning using high-order morphological brain networks derived from T1-w MRI for autism diagnosis. Front. Neuroinform. **12** (2018)

11. Dhifallah, S., Rekik, I.: Alzheimer's disease neuroimaging initiative and others: clustering-based multi-view network fusion for estimating brain network atlases of healthy and disordered populations. J. Neurosci. Methods **311**, 426–435 (2019)

12. Kipf, T.N., Welling, M.: Semi-supervised classification with graph convolutional networks. arXiv preprint arXiv:1609.02907 (2016)

13. Pan, S., Hu, R., Long, G., Jiang, J., Yao, L., Zhang, C.: Adversarially regularized graph autoencoder for graph embedding. arXiv preprint arXiv:1802.04407 (2018)

14. Mueller, S.G., et al.: The Alzheimer's disease neuroimaging initiative. Neuroimaging Clin. **15**, 869–877 (2005)

A Machine Learning Framework for Accurate Functional Connectome Fingerprinting and an Application of a Siamese Network

Ali Shojaee, Kendrick Li, and Gowtham Atluri$^{(\boxtimes)}$

Department of EECS, University of Cincinnati, Cincinnati, OH 45221, USA
shojaeai@mail.uc.edu, likt@mail.uc.edu, atlurigm@ucmail.uc.edu

Abstract. The goal of functional connectome (FC) fingerprinting is to uniquely identify subjects based on their functional connectome. In recent years, interest in this problem has increased substantially with efforts made to understand the factors that affect the accuracy of fingerprinting and to develop more effective approaches. In this work, we developed a novel machine learning framework for FC fingerprinting. Specifically, while existing approaches match a query FC with a reference FC based on a correlation score between the two FCs, our framework employed a machine learning model to determine if two FCs are similar. This allowed us to capture more complex features from FCs and also to capture non-linear similarities that may exist among FCs. We explored multiple machine learning algorithms that include a Siamese neural network and several classification algorithms. From our experiments, we observed that the Siamese network outperformed other classification models, with an FC fingerprinting accuracy of 99.89%.

Keywords: Functional connectivity · Fingerprinting · Parcellation · Precision neuroscience

1 Introduction

Functional connectivity (FC) that captures similarity in the blood oxygen level dependent (BOLD) signal, measured using functional Magnetic Resonance Imaging (fMRI), in different brain regions has emerged as one of the popular frameworks for analyzing fMRI data [4,5,8,13]. Despite the success of the FC framework in discovering key principles of brain function and dysfunction consistently at a group level [7,20,23], a major criticism is that there has been little success in discovering subject-specific principles and in explaining the neural basis of individual behavior [10,19]. A research direction that has emerged recently to

---

A. Shojaee and K. Li—Contributed equally to this paper.

© Springer Nature Switzerland AG 2019
M. D. Schirmer et al. (Eds.): CNI 2019, LNCS 11848, pp. 83–94, 2019.
https://doi.org/10.1007/978-3-030-32391-2_9

address this concern is *functional connectivity fingerprinting*, where the goal is to uniquely identify individuals using their FC [11]. Specifically, given a set of n reference FCs, one from each subject, and an unlabeled 'query' FC from one of these subjects, the goal of FC fingerprinting is to match the query FC with the reference FC that is also obtained from the same subject.

In the recent years, this problem of FC fingerprinting has gained significant attention in the neuroimaging community with numerous efforts to quantify FC accuracy on different datasets [2,3,11,17,24], to study the impact of different factors (e.g., sample size and granularity) on accuracy [16], to identify elements of the FC that result in better accuracy [11], and to develop better approaches for effective fingerprinting [2,3,17]. However, the common framework used by all these methods involve identification of the matching FC based on correlation between the query FC and the reference FCs. Correlation captures a linear relationship between two data objects and is not suited to capture similarity in higher-order transformations of features or to capture any non-linear relationship that may exist.

In this study we developed a machine learning framework for FC fingerprinting where we replaced the correlation-based matching with a machine learning (ML) model to determine if two FCs (a query FC and a reference FC) belong to the same subject. In doing so our framework leverages the advances in machine learning models to capture more complex scenarios that correlation-based matching is not designed for. Moreover, we used a neural network framework called a Siamese network, that is capable of learning models to automatically capture similarity and dissimilarity between objects, to accurately perform FC fingerprinting. We studied the performance of our Siamese network based fingerprinting approach using fMRI data acquired from the 1200-subjects March 2017 human connectome project (HCP) data release [22]. In addition to using a Siamese network, we also explored the utility of traditional classification algorithms such as k-nearest-neighbor (kNN), decision trees (DT), and Naive Bayes algorithm. We observed that our Siamese network outperformed other classification models, with an FC fingerprinting accuracy of 99.89%.

The contributions of this work to advances in connectomics in neuorimaging are multi-fold: 1. A suitable machine learning framework for FC fingerprinting that can employ any of the traditional classification algorithms; 2. Use of Siamese network, a neural network model, that is traditionally used for computing similarity and dissimilarity between data objects by automatically constructing the necessary higher-order features; 3. A systematic comparison of the performance of our Siamese network with traditional classification models in our machine learning framework.

The rest of the paper is structured as follows. The datasets used in this study and the preprocessing steps involved are presented in Sect. 2. An introduction to our new ML framework, the Siamese network and our experimental setup are discussed in Sect. 3. Our results are discussed in Sect. 4. Concluding remarks are presented in Sect. 5.

2   Data

The fMRI data used in this study was acquired from the 1200-subjects March 2017 human connectome project (HCP) data release [22] which included preprocessed resting state fMRI scans from 1003 healthy young adults (ages 22–35, 469 Male, 534 Female). Under the HCP protocol, all participants provided written informed consent, and the HCP was approved by the Institutional Review board at Washington University in St. Louis.

As part of the HCP, four resting-state fMRI scans were collected from each subject over two days. Two approximately 15 min scans were obtained each day, a left-to-right (LR) phase-encoded scan and a right-to-left (RL) phase-encoded scan. For our study, we used the extensively-preprocessed node-timeseries data made available in the HCP data release. This node-timeseries data was generated through the HCP preprocessing pipeline which included fMRI preprocessing, artefact removal using independent component analysis (ICA), inter-subject registration, group principal component analysis (PCA), parcellation through group-ICA, and dual-regression to compute the time series for each independent component (IC). Details of these steps are provided in the HCP documentation [1]. As part of the group-ICA step of this pipeline, different number of ICs have been derived and as a result node-timeseries are available at different granularities: 15, 25, 50, 100, 200, and 300. Based on prior studies which showed improved FC fingerprinting performance at finer parcellation granularity, we used the per-scan node-timeseries from the 300 ICs [2,16,17]. This data was used as is without further processing.

3   Methods

3.1   FC Generation

To construct FCs from each subject's fMRI scan, we computed the Pearson correlation for each pair of node timeseries and generated a pairwise correlation matrix. This resulted in 4012 FCs, four for each subject, i.e., one for each scan.

3.2   ML Framework for FC Fingerprinting

Unlike the traditional framework for FC fingerprinting, where the reference FC with the highest correlation with the query FC is treated as a match, we propose to use machine learning models to determine if two given FCs are from the same subject. To this end, we first created a training dataset, where a pair of FCs $[FC_a \ FC_b]$ is treated as one training instance, with a label 0 if they belong to the same subject and 1 otherwise. We ensured that our training set is balanced with similar number of 0 instances and 1 instances. We then trained machine learning models, including Siamese networks (described in Sect. 3.3) and other traditional classification models (described in Sect. 3.6). We then tested our models on test instances that comprised of pairs of FCs to predict if each pair is from the same subject or not. 5-fold cross-validation accuracy is computed to assess the performance of our models. The advantage of this ML framework over the traditional approach is that it leverages machine learning algorithms to learn the best possible model for fingerprinting.

3.3 Siamese Networks

A Siamese network is a neural network model which allows for the automatic computation of high-level features optimized for computing similarity or dissimilarity between two objects (e.g., images or signatures). Siamese network was originally designed for signature verification [6], later found applications in image retrieval [18] and more recently for predicting mental illness from an fMRI scan [14].

A Siamese network is comprised of two components: (1) the *feature engineering component* that learns the high-level features critical for assessing similarity or dissimilarity, and (2) the *similarity-computation component* where similarity is computed based on the features engineered by the first component.

In the original Siamese network that takes images as input, the feature engineering component consists of two convolutional neural networks (CNNs) with linked, identical weights where any change to one CNN is mirrored on the other. CNNs are known for their ability to automatically learn features from images for a variety of image processing tasks. To determine whether two input images are from the same class, one input is fed into one CNN and the other into the other CNN. Mirrored CNNs allow the same features to be extracted from the two input images separately. The output features from both CNNs are then fed into the similarity-computation component. The similarity-computation component can consist of any function which can compute similarity between two vectors (e.g., an L_2 norm) or with the help of more complex models (e.g., a fully connected neural network).

With an appropriate loss function, this architecture trains a single CNN to produce high-level features which are maximally different between classes and minimally different within classes. As only a single CNN is trained to produce high-level features and the CNN is uncoupled from the similarity component, a Siamese network can provide additional flexibility by allowing the data to be pre-processed into high-level features in parallel before the similarity-computation component and simplifies the prediction pipeline by being pair-order-independent (i.e., $[FC_a \ FC_b]$ and $[FC_b \ FC_a]$ are indistinguishable from one another).

Graph Convolution Networks (GCNs): In a traditional convolutional network, localized features are learned as convolutional filters from 2D or 3D images. These convolutional filters are activated when the values in a local region of the image match with that of the filter, signalling that a desired feature is found in the image. In this case, pixels have predefined relationships with neighboring pixels, and such spatial relationships are consistent throughout the image, i.e., each pixel is adjacent to its neighbors.

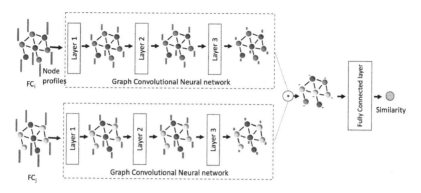

Fig. 1. Siamese network with a graph convolution network for feature engineering and a fully connected layer for similarity computation. This is the network architecture we used for fingerprinting in this paper.

In our case, we are interested in FC fingerprinting, i.e, computing similarity between graphs. Graphs lack the regular structure that exist among pixels as each node may have differing number of neighbors. Due to this difference, one must first redefine the convolutional operation in the graph setting. Shuman et al. [21] defined filters in the graph spectral domain. Utilizing the normalized graph Laplacian (defined as $L = I_R - D^{1/2}AD^{1/2}$ where I_R is the identity matrix, D is the diagonal degree matrix and A is the adjacency matrix), L can be decomposed as $L = U\Lambda U$, where U is the matrix of eigenvectors and Λ is the diagonal matrix of eigenvalues. The graph Fourier transform of some signal $\mathbf{c}$ can then be expressed as $\mathbf{c} = U^T\mathbf{c}$. These two properties allow us to define the convolution on a graph in the spectral domain as $g_\theta \circledast \mathbf{c} = U g_\theta U^T \mathbf{c}$, where $g_\theta = diag(\theta)$ and θ is a vector of Fourier coefficients. As this method can result in excessive computations and produce spectral filters outside the graph spatial domain, Hammond et al. [12] proposed a polynomial parametrization of the filters directly on the Laplacian. Defferrard et al. [9] then proposed approximating these polynomials as Chebyshev polynomials, reducing computational complexity. Filtering signal $\mathbf{c}$ with a K localized filter is then performed by:

$$y = g_\theta(L) \circledast \mathbf{c} = \sum_{k=0}^{K} \theta_k T_k(\tilde{L})\mathbf{c}$$

where $\tilde{L} = (2/\lambda_{max})L * I_R$, λ_{max} is the largest eigenvalue of L, and $T_k(c)$ is the kth Chebyshev polynomial for c. The output for the jth layer for a sample s in a GCN is thus:

$$y_{s,j} = \sum_{i=1}^{F_i n} g_{\theta i,j}(L)c_{s,i} \in \mathbb{R}^R$$

where $c_{s,i}$ is the input feature map. We use Defferrard et al.'s [9] version of GCNs in this paper.

Structure of the Siamese GCN: In this work, we use a Siamese GCN that is similar to the one proposed by Ktena et al. [14]. Two GCNs with linked, identical weights are fed into a fully connected neural network layer which computes the distance between the two input graphs. This is shown in Fig. 1. Within each GCN, each convolutional layer contains a rectified linear unit (ReLU) activation, allowing for non-linear transformations when determining high-level features.

Defining Loss for Training: The global loss was proposed by Kumar et al. [15] to improve robustness to outliers and to provide better regularization than the commonly used hinge loss. The objective of the global loss is to maximize the distance between the means of the two classes (0 and 1) while minimizing the variance for the two classes to effectively distinguish between the two classes. The equation for the global loss is:

$$J^{global} = (\sigma_0^2 + \sigma_1^2) + \lambda max(0, m - (\mu_0 - \mu_1))$$

where σ_0^2, σ_1^2 are the variances, μ_0, μ_1 are the means of matching and non-matching pairs, respectively, m is the margin between the output values of matching and non-matching pairs, and λ is a scaling parameter between variance minimization or mean difference maximization.

3.4   Graph Generation

While an FC is a graph with nodes and edges, FC pairs $[FC_a \ FC_b]$ cannot be directly fed into a Siamese network with a GCN. For the GCN we defined above, a graph needs to have attributes for each node. Another choice that can influence the GCN is the choice of edges in the FC. Here we describe our approach to transforming an FC into a graph that can be passed on to a GCN in our Siamese network.

First, the attributes for a node is defined as the FC profile of that region, that is, the vector of correlations between that region's BOLD time series and all other regions' BOLD time series. This can be extracted from the FC simply by taking all elements in that region's row or column, excluding the diagonal value.

Secondly, we defined the edges between nodes. One simple approach we used is a fully connected graph where all nodes are connected to all other nodes. We refer to this as the *Complete graph*. For the complete graph, any node can be directly influenced by any other node. Alternatively, one can define neighboring relationships selectively. In our case, we computed the group average FC from all FCs to capture the common functional relationships among samples. From the group average FC, we removed edges with the lowest magnitude correlation from the fully connected graph until we could no longer remove an edge without disconnecting the graph. By doing so, we created a connected graph which retains only the strongest functionally connected edges. We refer to this as the *Functional graph*. In contrast to the complete graph, by removing weak edges we introduced functional locality to the graph, where only nodes functionally connected can directly influence each other.

3.5   Siamese GCN Implementation

Same-subject and different-subject pairs were generated for our experiments by pairing different FCs with each other, resulting in 6018 $(1003 * \binom{4}{2})$ same-subject pairs and $8,040,048$ $(\binom{1003*4}{2}) - 6018)$ different-subject pairs. Of these different-subject pairs, 6000 were randomly selected to create a balanced $12,018$ FC-pair dataset. In this dataset, same-subject and different-subject pairs were labeled 0 and 1, respectively. We evaluated the performance of the Siamese GCN model using 5-fold cross validation. The model was trained using two different graph structures: (1) the complete graph and (2) the functional graph, both described previously in Sect. 3.4.

Two graphs and their feature vectors are passed separately to each of the two GCNs that comprise our *feature engineering component*, where each graph convolutional layer is followed by a rectified linear unit (ReLU) activation function to capture non-linearity. Following the GCNs, we compute the inner product with a dropout of 20%. This dropout randomly sets the specified percent of nodes to zero. After this dropout, the output is then passed to the similarity learning network, a fully connected layer with linear activation that constitutes our *similarity computation component*. The output layer of this fully connected layer returns the similarity of the two input graphs which is then used to classify the pair. For training, we used the global loss function to calculate loss and optimize weights.

We used different GCN architectures for the complete graph and the functional graph. For the complete graph, we used a two layer GCN each with 64 features. We refer to this configuration as the *Complete_64_64* network. For the functional graph, we used two different GCN architectures: 1.*Functional_64_64*, 2.*Functional_64_16_4*. *Functional_64_64* is identical in structure to *Complete_64_64*. For *Functional_64_16_4*, we used a three layer GCN with feature sizes of 64, 16, and 4 for each layer respectively.

All networks were optimized using stochastic gradient descent with Adaptive Moment Estimation (ADAM) optimizer. For training, we used a learning rate of 0.001, a regularization parameter of 0.005, and global loss function parameter values of $\lambda = 0.35$ and margin $m = 0.6$. We also used a filter order parameter of $k = 3$, which allows filters to use node features from neighbors 3 hops away.

3.6   Traditional Classification Techniques

As part of our proposed ML framework for fingerprinting we used traditional classification techniques such as k-nearest neighbor (kNN), decision trees (DT), and Naive Bayes (NB). To train these models, we first generated vectorized FCs by taking upper triangular values of the FC matrix. Same-subject and different-subject pairs were then generated by concatenating vectorized FCs. Once vectorized and concatenated, the FC-pair database was generated similarly as described above for the Siamese GCN.

Our Siamese network setup was pair-order-independent, i.e., pairs $[FC_a \ FC_b]$ and $[FC_b \ FC_a]$ are treated as the same. Whereas in the case of classification, each element of the concatenated vectorized FCs will be treated as features and so an FC pair $[FC_a\ FC_b]$ will be treated differently from $[FC_b\ FC_a]$. To address this issue, for every pair $[FC_a\ FC_b]$ in the dataset, we also added the pair $[FC_b\ FC_a]$ with the same label to the training dataset.

We evaluated the performance of kNN, DT, and NB classification for FC fingerprinting by performing 5-fold cross validation as previously described for our Siamese network. We used MATLAB's *fitcknn* and *fitctree* to learn kNN and DT classification models for k $= 1, 3, 5$ and MaxNumSplits $= 5, 10, 15, unconstrained$ respectively using Euclidean distance. For our NB evaluation, we used scikit-learn's *GaussianNB* to learn NB classification models.

4   Results

4.1   FC Fingerprinting Performance

We performed a comparative analysis of FC fingerprinting performance of our ML framework with a Siamese network, kNN classification, DT classification, and a Naive Bayes classification. The results for our comparative analysis are shown in Table 1. We observed that the Siamese GCN performed extremely well compared to all other methods, with the best performing Siamese GCN *Functional_64_16_4* achieving 99.89% accuracy. The worst performing Siamese GCN is the *Complete_64_64*, achieving 99.56% prediction accuracy. The reason for the superior (nearly 100% accuracy) performance of the Siamese GCNs is that the network is able to construct suitable higher-order features that are highly relevant for fingerprinting. We observed that the *Complete* model performed marginally weaker than *Functional* models, which suggest that by limiting the interactions between nodes to a functionally local neighborhood, the Siamese GCN was marginally better at extracting key subject-specific information. Furthermore, the *Functional_64_16_4* model was able to perform better than the *Functional_64_64* model which did not use shrinking layers. This observation suggests that the Siamese GCN is more able to produce better differentiating higher-order FC features when the deeper layers are constrained in the number of features.

Among the classification models, the best model, kNN, was only able to achieve 76.16%, where the closest neighbor (i.e., $k = 1$) was used to classify an FC pair. This configuration is very similar in flavor to traditional correlation-based FC fingerprinting, where FCs are labeled based on the highest correlating FC. The only difference is that in kNN we are predicting if a given pair of FCs are from the same subject, whereas in traditional correlation-based matching we are identifying the best matching subject based on the nearest match. Our DT and Naive Bayes models performed poorly, with the best DT model only achieving an accuracy of 55.51% that is close to what is expected of a random classifier and the Naive Bayes model achieving slightly below the expected accuracy of a random classifier with a 46.97% accuracy. One potential reason for the relatively poor performance of the classification techniques is the high dimensionality of the data, with $89,700$ features for each sample (i.e., an FC pair).

Table 1. Fingerprinting performance of our machine learning based fingerprinting framework.

Models	Parameters	Accuracy
Siamese GCN	Complete_64_64	**99.56%**
	Functional_64_64	**99.73%**
	Functional_64_16_4	**99.89%**
kNN	k=1	76.16%
	k = 3	68.55%
	k = 5	60.23%
DT	Max nodes = 5	54.30%
	Max nodes = 10	54.76%
	Max nodes = 15	54.57%
	No maximum	55.51%
Naive Bayes	-	46.97%

4.2 Exploring the Similarity Computation Component of Siamese Network

One of the criticisms against neural networks is that they are limited in providing interpretable models, particularly in the case of GCNs where higher-order features are constructed. However, the fully connected layer within our similarity computation component (Fig. 1) can be studied to determine which region's higher-order features are weighted more for computing the similarity. The weights on the regions from the fully connected layer are shown in Fig. 2. From this figure, it can be observed that the regions used by the fully connected layer are symmetric across the two hemispheres. Specifically, para-central, cuneus, superior frontal regions appear to have been preferred for assessing similarity between FCs.

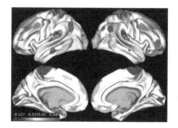

Fig. 2. Brain maps showing the weights for each region in the fully connected layer of the Siamese network.

Shown in Fig. 3 is the distribution of similarity scores estimated by Siamese GCN *Functional_64_16_4* model for all matching and non-matching pairs in our

Table 2. Confusion matrices for the (a) Complete_64_64, (b) Functional_64_64, (c) Functional_64_16_4 model from the 5-fold cross validation on the full FC pair dataset.

(a)

	Predicted Label	
True Label	0	1
0	5991	27
1	26	5974

(b)

	Predicted Label	
True Label	0	1
0	6002	16
1	17	5983

(c)

	Predicted Label	
True Label	0	1
0	6012	6
1	7	5993

FC-pair dataset. From this figure, it can be observed that the two classes are well separated by a similarity threshold of approximately -1.0. This suggests that the Siamese GCN is able learn the key higher-order features from FC matrices that reflect similarity between FCs from the same subject and further explains the reason for Siamese GCN's superior and nearly 100% fingerprinting accuracy.

To further illustrate the superior performance of the Siamese GCNs, the confusion matrices from 5-fold cross validation experiment are shown in Table 2. Here 0 indicates pairs of FCs from the same subject, and 1 indicates pairs of FCs from different subjects. In line with the above reported fingerprinting performance results, we observed very few false negatives (6 compared to 16, 27) and false positives (7 compared to 17, 27) for the best performing Siamese GCN model *Functional_64_16_4* compared to *Functional_64_64* and *Complete_64_64* respectively. We observed that the number of false positives was very similar to the number of false negatives for all models.

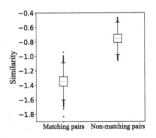

Fig. 3. Box plots of similarity scores for matching and non-matching pairs.

In summary, the Siamese GCNs resulted in nearly 100% fingerprinting accuracy owing to their ability to learn relevant higher-order features from the training data.

5   Conclusion

In this work, we introduced a new ML framework for effective FC fingerprinting. In the context of this framework, we investigated the potential of Siamese

GCNs in learning higher order FC graph features for effective FC fingerprinting. We evaluated several models, including classification models and Siamese network with different configurations of GCNs. We observed that Siamese networks outperformed traditional classification methods and achieved 99.89% FC fingerprinting accuracy. While the GCN Siamese network was able to effectively engineer the relevant higher-order features suited for fingerprinting, traditional classification models were not able to handle the problem of high-dimensionality, due to which many of these models resulted in an accuracy closer to that of a random classifier.

While this work has uncovered various insights and observations previously unexplored in the context of FC fingerprinting, there are several more aspects which can be explored in more detail. One area of interest is in exploring the different parameter settings, such as the number of layers, and their effect on finding effective higher-order features of the FC. Another area of interest is the computational complexity of GCN learning. Despite using the state-of-the-art algorithms, GCN training for our experiments was slow, where each fold required approximately 3–4 h of learning on a state-of-the-art desktop computer with a 3.70 GHz AMD Ryzen 7 2700X eight core processor and 16 GB of memory. Additionally, there have been many studies on the effect of the granularity of parcellation and number of subjects on FC fingerprinting accuracy [2,16,17,24]. The impact the size of the training set or the granularity of parcellation has on Siamese GCN FC fingerprinting performance could be investigated.

Acknowledgements. This work was supported by NSF Grant IIS-1850204. The computational work is performed using the Data Analytics Cluster acquired through the Ohio Dept. of Higher Education's RAPIDS grant in 2018.

References

1. HCP documentation. https://www.humanconnectome.org/storage/app/media/documentation/s1200/HCP1200-DenseConnectome+PTN+Appendix-July2017.pdf
2. Airan, R.D., Vogelstein, J.T., Pillai, J.J., Caffo, B., Pekar, J.J., Sair, H.I.: Factors affecting characterization and localization of inter-individual differences in functional connectivity using MRI. Hum. Brain Mapp. **37**(5), 1986–1997 (2016)
3. Amico, E., Goñi, J.: The quest for identifiability in human functional connectomes. Sci. Rep. **8**(1), 8254 (2018)
4. Atluri, G., MacDonald III, A., Lim, K.O., Kumar, V.: The brain-network paradigm: using functional imaging data to study how the brain works. Computer **49**(10), 65–71 (2016)
5. Bargmann, C.I., Marder, E.: From the connectome to brain function. Nat. Methods **10**(6), 483 (2013)
6. Bromley, J., Guyon, I., LeCun, Y., Säckinger, E., Shah, R.: Signature verification using a "siamese" time delay neural network. In: Advances in Neural Information Processing Systems, pp. 737–744 (1994)
7. Bullmore, E., Sporns, O.: The economy of brain network organization. Nat. Rev. Neurosci. **13**(5), 336 (2012)

8. Castellanos, F.X., Di Martino, A., Craddock, R.C., Mehta, A.D., Milham, M.P.: Clinical applications of the functional connectome. Neuroimage **80**, 527–540 (2013)

9. Defferrard, M., Bresson, X., Vandergheynst, P.: Convolutional neural networks on graphs with fast localized spectral filtering. In: Advances in Neural Information Processing Systems, pp. 3844–3852 (2016)

10. Dubois, J., Adolphs, R.: Building a science of individual differences from fMRI. Trends Cogn. Sci. **20**(6), 425–443 (2016)

11. Finn, E.S., et al.: Functional connectome fingerprinting: identifying individuals using patterns of brain connectivity. Nat. Neurosci. **18**(11), 1664 (2015)

12. Hammond, D.K., Vandergheynst, P., Gribonval, R.: Wavelets on graphs via spectral graph theory. Appl. Comput. Harmon. Anal. **30**(2), 129–150 (2011)

13. Kelly, C., Biswal, B.B., Craddock, R.C., Castellanos, F.X., Milham, M.P.: Characterizing variation in the functional connectome: promise and pitfalls. Trends Cogn. Sci. **16**(3), 181–188 (2012)

14. Ktena, S.I., et al.: Distance metric learning using graph convolutional networks: application to functional brain networks. In: Descoteaux, M., Maier-Hein, L., Franz, A., Jannin, P., Collins, D.L., Duchesne, S. (eds.) MICCAI 2017. LNCS, vol. 10433, pp. 469–477. Springer, Cham (2017). https://doi.org/10.1007/978-3-319-66182-7_54

15. Kumar, B., Carneiro, G., Reid, I., et al.: Learning local image descriptors with deep siamese and triplet convolutional networks by minimising global loss functions. In: Proceedings of the IEEE Conference on Computer Vision and Pattern Recognition, pp. 5385–5394 (2016)

16. Li, K., Atluri, G.: Towards effective functional connectome fingerprinting. In: Wu, G., Rekik, I., Schirmer, M.D., Chung, A.W., Munsell, B. (eds.) CNI 2018. LNCS, vol. 11083, pp. 107–116. Springer, Cham (2018). https://doi.org/10.1007/978-3-030-00755-3_12

17. Peña-Gómez, C., Avena-Koenigsberger, A., Sepulcre, J., Sporns, O.: Spatiotemporal network markers of individual variability in the human functional connectome. Cereb. Cortex **28**, 2922–2934 (2017)

18. Qi, Y., Song, Y.Z., Zhang, H., Liu, J.: Sketch-based image retrieval via siamese convolutional neural network. In: 2016 IEEE International Conference on Image Processing (ICIP), pp. 2460–2464. IEEE (2016)

19. Rosen, B.R., Savoy, R.L.: fMRI at 20: has it changed the world? Neuroimage **62**(2), 1316–1324 (2012)

20. Shehzad, Z., et al.: The resting brain: unconstrained yet reliable. Cereb. Cortex **19**(10), 2209–2229 (2009)

21. Shuman, D.I., Narang, S.K., Frossard, P., Ortega, A., Vandergheynst, P.: The emerging field of signal processing on graphs: extending high-dimensional data analysis to networks and other irregular domains. arXiv preprint arXiv:1211.0053 (2012)

22. Smith, S.M., et al.: Resting-state fMRI in the human connectome project. Neuroimage **80**, 144–168 (2013)

23. Sporns, O.: The human connectome: a complex network. Ann. N. Y. Acad. Sci. **1224**(1), 109–125 (2011)

24. Waller, L., et al.: Evaluating the replicability, specificity, and generalizability of connectome fingerprints. Neuroimage **158**, 371–377 (2017)

Test-Retest Reliability of Functional Networks for Evaluation of Data-Driven Parcellation

Jianfeng Zeng, Anh The Dang, and Gowtham Atluri$^{(\boxtimes)}$

Department of EECS, University of Cincinnati, Cincinnati, OH 45221, USA
zengjg@mail.uc.edu, anhdt@mail.uc.edu, atlurigm@ucmail.uc.edu

Abstract. Brain parcellations play a key role in functional connectomics. A set of standard neuro-anatomical brain atlases are in common use in most studies. In addition, data-driven parcellations computed from fMRI data using a variety of clustering algorithms have also been used. Recent studies set out to determine the best parcellation in terms of quality and reliability have remained inconclusive without a clear winner. In this work, we investigated the utility of test-retest reliability of functional connectivity as an evaluation metric for comparing parcellations. Specifically, using data from the human connectome project, we compared a data-driven parcellation and a geometric parcellation using Intraclass Correlation Coefficient (ICC). We also investigated the impact of parcellation granularity on the test-retest reliability. We observed that the ICCs for geometric parcellation are better than those of a data-driven parcellation, suggesting that the FCs computed using regular parcels in the geometric atlases are more reliable than those computed using a data-driven parcellation.

Keywords: Functional connectivity · Test-Retest reliability · Parcellation · Precision neuroscience

1 Introduction

Functional Magnetic Resonance Imaging (fMRI) technology invented in the early 1990s was expected to have a significant impact on the diagnosis and treatment of mental illness, in addition to having a scientific impact in enabling the discovery of the governing principles of brain function [25]. Over the years many extensive studies have been conducted [7,16,23,29], and data analysis tools such as functional network analysis [18], dynamic functional network analysis [24], and independent component analysis (ICA) [8] have been developed for discovering insights from large fMRI datasets. After 25 years, one of the major criticisms on

---

J. Zeng and A. T. Dang—Contributed equally to this paper.

M. D. Schirmer et al. (Eds.): CNI 2019, LNCS 11848, pp. 95–105, 2019.
https://doi.org/10.1007/978-3-030-32391-2_10

fMRI research is that only the group-level principles of brain function have been discovered by averaging the signal across the group, with little success in discovering subject-specific principles [13,25]. One direction that is pursued to address this concern is to quantify the test-retest reliability of measures computed using subject-level fMRI data [4,10,11,20,31]. Existing studies explored the reliability of: (i) ICA components [31], (ii) graph theoretic properties [4,10,11], and (iii) the default mode network [20].

On the other hand, a major quest in the neuroscience community since the early 1900s [30] has been to identify distinct cortical regions in the human brain, which is now conventionally referred to as a *brain parcellation* [14]. With increased availability of fMRI datasets, a variety of clustering approaches such as spectral clustering [12], Ward hierarchical clustering [9], k-means clustering [26], hierarchical Dirichlet process mixture models [17], and von Mises-Fisher distribution based clustering [27] have been used to group voxels with similar time series or connectivity profiles to discover brain parcellations. Brain parcellations play a key role in the *functional connectome* (FC) framework where functional networks are derived from region-level time series computed by averaging the voxel-level time series within each parcel or region [6]. Most studies either use a neuro-anatomical atlas [28] or one of the data-driven atlases [12,15,27], often without a consensus on the choice of parameters (e.g., granularity of parcellation). This makes it challenging to compare results across studies, as the regions implicated in different studies are defined as part of different brain parcellations. In search of an ideal parcellation, a recent study [5] extensively evaluated different parcellations in terms of their quality and reproducibility, and observed that there was no one parcellation that outperformed others in both these criteria.

Considering the fact that reliability of the subject-level FC is a key requirement towards delivering the promise of fMRI technology in delineating the principles of brain function in individual subjects and the basis for individual behavior, in this work we explored the role of test-retest reliability of FC for evaluating brain parcellations. To the best of our knowledge, this is the first study that explored the test-retest variability of FCs as a metric for evaluating brain parcellations. Specifically, we investigated the following two questions: 1. Does the parcellation method affect test-retest reliability of resting-state FC? 2. Does the granularity of parcellation affect reliability of resting-state FC?

To this end, we compared the test-retest reliability of FC constructed using a data-driven parcellation with that of a geometric parcellation [26]. Note that a geometric parcellation does not rely on fMRI data, but only groups voxels based on their coordinates. We used Ward clustering algorithm to derive a data-driven parcellation that clusters voxels with similar time series into parcels. We used Intraclass Correlation Coefficient (ICC) to compute the test-retest reliability between two FCs computed from the same subject. We used resting-state fMRI data from the 1200-subject data release (S1200) [1] of the Human Connectome Project (HCP) in this study.

We observed that the test-retest reliability for geometric parcellation that does not rely on fMRI data is always higher than that of a data-driven parcellation. We also observed that the ICC decreases as the granularity of the parcellation increases, and this is likely to be the result of low signal-to-noise-ratio.

This work contributes to the advances in connectomics in neuroimaging in multiple ways. First, we introduced test-retest reliablity of functional connectivity as an evaluation metric for comparing parcellations. Second, we empirically compared the reliability of FCs using a data-driven parcellation and a geometric parcellation that does not take into account the fMRI data. Third, we studied the impact of parcellation granularity on test-retest reliability.

The rest of this paper is organized as follows. The dataset used along with the pre-processing steps are described in Sect. 2. Methods for computing data-driven and geometric parcellations, along with the ICC for quantifying test-retest reliability are discussed in Sect. 3. Our results are presented in Sect. 4. Conclusions are provided in Sect. 5.

2 Data

In this study, we used fMRI data from the 1200-subject data release (S1200) [1] of the Human Connectome Project (HCP). This release includes resting state (rsfMRI) and task-related (tfMRI) data from 1,003 healthy subjects. fMRI data was collected from each subject during two sessions, with two 15 min scans per session. This data was minimally processed using a number of steps including gradient distortion correction, motion detection, field map preprocessing, spline resampling, intensity normalization, and mapping timeseries from volume to CIFTI grayordinates standard space. More information about the data and the preprocessing steps is available in the S1200 data release manual [2].

We restricted our analysis to rsfMRI data from 50 unrelated subjects to avoid the effect of familial relationships on our analysis. For these subjects, we further restricted our analysis to the two fMRI scans from the first session. The acquired data was further processed using the following steps: 1. Band-pass filtering, 2. Global signal regression, and 3. Normalization. Band-pass filtering with the range [0.08 Hz 0.009 Hz] was performed to avoid the effect of non-neural dynamics on our analysis. Global signal regression [22] was performed to remove the global signal that is present in all or most of the voxels. Normalization was performed to ensure that voxels with high magnitude of BOLD signal do not dominate the computation of the region-level time series computed as the average of the voxel time series within a region. At the end of this normalization step, all the voxel time series have a zero mean and unit variance.

3 Methods

3.1 Computing Parcellations

In this section, we describe the two different types of parcellation we evaluated in our study: 1. A data-driven parcellation, 2. Geometric parcellation. The data-driven parcellation uses fMRI data to group voxels with highly similar Blood

Oxygen Level Dependent (BOLD) time series. For grouping similar voxels, we used Ward clustering which is a hierarchical clustering technique. For geometric parcellation, we grouped voxels whose locations are in close proximity. This merely groups contiguous locations into parcels. We used k-means algorithm for geometric parcellation.

Note that, because Ward-based parcellation depends on BOLD signal, each subject has a specific version of Ward parcellation. On the other hand, the geometric parcellations are independent of BOLD signal and so they are not subject-specific. The parcellations were computed at different granularities so we can study the impact of granularity on test-retest reliability. The granularities used are 200, 400, 600, 800, 1000, 2000, and 4000.

Below we describe the setup used for each of the two types of parcellation.

Ward-Based Parcellation. To compute Ward-based parcellation for each subject, we first temporally concatenated the two rs-fMRI scans from the first session. The voxels in the left and right hemisphere were extracted, and a pairwise distance matrix for all the voxels in each hemisphere was computed using Euclidean distance. The pairwise distance matrices were then separately passed to the Ward's clustering algorithm implemented in Julia language's clustering package [3]. Note that the spatial contiguity was not enforced during Ward clustering, and so the voxels in resultant parcels are not necessarily bound to be contiguous.

Geometric Parcellation. To compute a geometric parcellation, we first extracted voxel coordinates from a standard-template surface files for the two hemispheres. These voxel coordinates were used as features to derive the geometric parcellation using the k-means algorithm implemented in Julia language's clustering package [3]. We used Euclidean distance metric for k-means clustering. Because the parcellations are only based on voxel locations, these parcels are more regular in size compared to the Ward parcellation. Moreover, we chose k-means clustering because we are likely to get a different parcellation on every instantiation of the k-means algorithm. To assess the consistency of test-retest reliability, we constructed and used 10 different Geometric parcellations for each resolution.

3.2 Computing Test-Retest Reliability Using ICC

Test-retest reliability indicates the extent to which the measurements can be replicated in two different instances. In our case, the measurements are the values (correlations) in the FC matrix and the two instances are two rsfMRI scans. The rest-retest reliability of the FC matrix was computed using Intraclass Correlation Coefficient (ICC) [19]. Of the 10 different versions of ICC that are available for quantifying replicability of measurements, we selected the 'Two-way mixed effects' model, with 'single measurement' type, and a definition of 'absolute agreement', based on the flow-chart provided in [19]. This is because

(1) we used all elements of the FC matrix that were computed from all two scans (two-way mixed effects), (2) values of each FC matrix in two scans were used separately instead of their across-scan mean (single measurement), and (3) the across-scan absolute agreement was of interest to us. This version of ICC is also referred to as ICC(2,1) (as in [10]), and the formula was defined in [21] as follows:

$$ICC(2,1) = \frac{BMS - EMS}{BMS + (k-1)EMS + \frac{k}{n}(JMS - EMS)} \tag{1}$$

where BMS is the between-targets (scans) mean square, EMS is the error mean square, JMS is the between-subjects (edges in the FC) mean square, k is the number of targets (scans), and n is the number of subjects. Interested readers are referred to [21] for more information on the different versions of ICC and to [19] for recommendations on which versions to use for different experimental settings.

4   Results

4.1   Impact of Parcellation on Test-Retest Reliability

Results on One Subject: The ICC scores computed for data from one subject 100307 using Ward parcellation and 10 different geometric parcellations at different granularities are shown in Fig. 1(a). It can be observed from the figure that the ICC scores computed using the Ward parcellation for all granularities are lower compared to those of each of the 10 geometric parcellations. The brain maps for the Ward and 10 geomertic parcellations are shown in Fig. 1(b). From this figure, it can be observed that the geometric parcellations are more regularly/circularly shaped compared to the Ward parcellation with arbitrary shapes. Also, note that the geometric parcellations are oblivious to the bold signal. While Fig. 1 shows the ICC scores for only one subject, this is a trend that we observed over all the 50 subjects used in this study. Some evidence for this can also be seen in Figs. 3 and 4 as discussed below.

It is also interesting to note from Fig. 1(a) that the ICCs for higher granularities are lower than those of lower granularities for both Ward and geometric parcellations. This suggests that the FCs computed using coarser parcellations are more reliable.

A comparison of the edges in FCs for subject 100307 computed from scan 1 and scan 2 is shown in Fig. 2 for Ward and geometric clustering separately. These plots are showing densities in 2D. The ICC score for the FCs computed using Ward clustering is 0.44 and that of geometric clustering shown in this figure is 0.55. While it is not markedly evident from the figure, the plot for geometric parcellation is more elliptical compared to that of the Ward parcellation. As a result the reliability for Ward clustering is somewhat weaker compared to that of Geometric clustering. This indicates that the network edges between two scans computed using Geometric parcellation are more in agreement compared to the case of Ward parcellation.

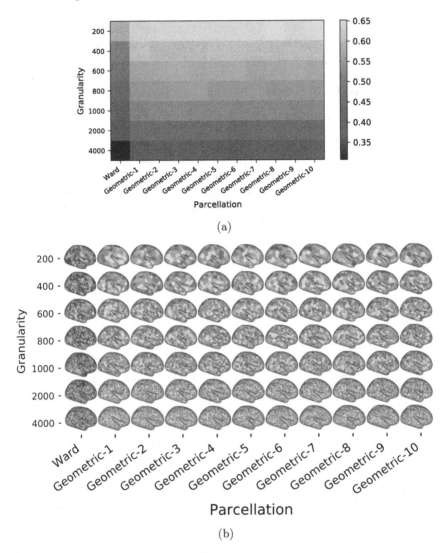

Fig. 1. (a) Test-retest reliability of FCs computed using data from subject 100307, using Ward and Geometric parcellations at different granularities. (b) Brain maps of the Ward and Geometric parcellations computed at different granularities (best viewed in color and in a magnified version). (Color figure online)

Results on Multiple Subjects: To study the generality of our above observations, we present the ICC scores computed using data from 10 subjects in Fig. 3. The figure shows ICC scores of FCs computed using Ward parcellation at a granularity of 800 regions in red and the boxplot of ICC scores for geometric parcellation in blue. It can be observed from the Fig. 3 that the ICC scores for FCs constructed using geometric parcellation are superior to that of Ward

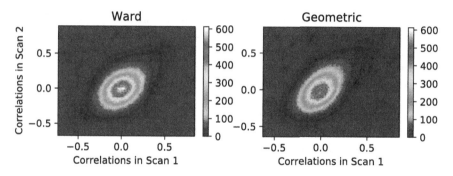

Fig. 2. 2D histograms showing the comparison between FCs in Scan 1 and Scan 2 computed using Ward clustering (*left*) and geometric clustering (*right*) at the granularity of 800 regions. The ICC score for the Ward clustering on the left is 0.44. The ICC score for the geometric clustering on the right is 0.55.

parcellation, and are consistent across all 10 subjects. It is also interesting to note that despite the large variability in ICC scores across subjects for each type of parcellation, the ICC scores for FCs constructed using geometric parcellation are higher.

The distribution of ICC scores computed for FCs from all 50 subjects using Ward parcellation and one of the 10 geometric parcellations at the granularity of 400, 600, 800, and 1000 are shown in Fig. 4. It can be observed that the means of

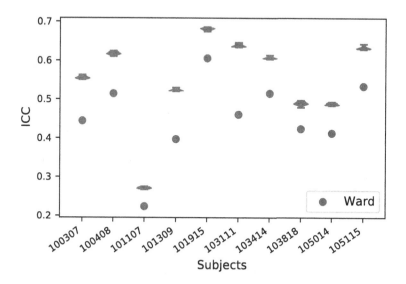

Fig. 3. Comparison of ICC scores between Ward clustering (red) and geometric clustering (box plot in blue) for 10 subjects when a parcellation granularity of 800 regions is used. (Color figure online)

the distributions of Ward parcellation for all granularities are *significantly* lower than the ones of the geometric parcellation. This further illustrates that across all 50 subjects, the ICC scores computed using Ward parcellation at different granularities are lower compared to those of the geometric parcellation. The reason for overlap among the two distributions in Fig. 4 is the high variability in ICC scores across subjects as seen in Fig. 3.

In summary, geometric parcellations that are indepedent of the data and have more regular shapes, as seen in Fig. 1(b), result in more reliable FCs than Ward parcellations that are data-driven.

4.2   Impact of Granularity of Parcellation on Test-Retest Reliability

To study the impact of granularity of parcellation on ICC scores, we compared the ICC scores computed for data from subject 100307 using Ward parcellation and 10 geometric parcellations for different granularities in Fig. 5. Not only does the plot indicate that ICC scores computed using Ward parcellation are all lower than the ICC scores computed using 10 geometric parcellations at each granularity, but it also reinforces the observation we made in Fig. 1 that ICC scores for higher granularities are lower than those of lower granularities for both Ward and geometric parcellations. One reason for this decrease in ICC with increase in granularity is likely to be the low signal-to-noise ratio with only small number of voxels present in each region (approximately 10 voxels/region, on average).

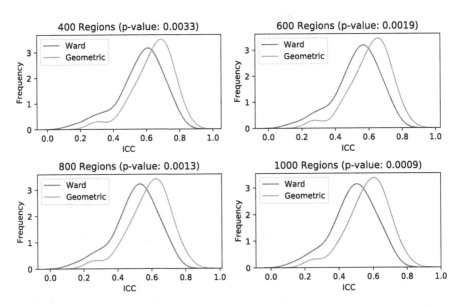

Fig. 4. Comparison of distribution of ICC scores when Ward and geometric parcellation are used. Each of these distribution have ICC scores from 50 subjects. Each subplot is for one granularity.

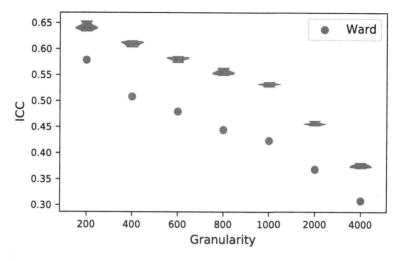

Fig. 5. Comparison of ICC scores between Ward clustering (red) and geometric clustering (box plot in blue) for the subject 100307 for all granularities. (Color figure online)

5   Conclusion

In this work, we investigated the role of test-retest reliability as an evaluation metric for comparing different parcellations. Our results suggested that geometric parcellations with more regular-shaped parcels are likely to yield better reliability in FCs than data-driven parcellations with arbitrarily shaped parcels. We observed that the reliability in FC decreases with increase in parcellation granularity. This is likely to be due to low signal-to-noise-ratio, as there are only a few voxels per region at high granularities.

One limitation of our study is that it was limited to one type of data-driven clustering (Ward clustering). As part of future work, one could perform a more extensive comparison of multiple data-driven parcellation approaches such a spectral clustering, average-link hierarchical clustering, and hierarchical Dirichlet process mixture models. Another limitation of this study is that it did not investigate the reasons for reduced ICC scores for data-driven parcellations, and it is a relevant topic for future research.

Acknowledgements. This work was supported by NSF Grant IIS-1850204. The computational work is performed using the Data Analytics Cluster acquired through the Ohio Dept. of Higher Education's RAPIDS grant in 2018.

References

1. HCP 1200 subjects data release. https://www.humanconnectome.org/study/hcp-young-adult/document/1200-subjects-data-release

2. HCP S1200 release reference manual. https://www.humanconnectome.org/storage/app/media/documentation/s1200/HCP_S1200_Release_Reference_Manual.pdf

3. Julia clustering package. https://github.com/JuliaStats/Clustering.jl

4. Andellini, M., Cannatà, V., Gazzellini, S., Bernardi, B., Napolitano, A.: Test-retest reliability of graph metrics of resting state MRI functional brain networks: a review. J. Neurosci. Methods **253**, 183–192 (2015)

5. Arslan, S., Ktena, S.I., Makropoulos, A., Robinson, E.C., Rueckert, D., Parisot, S.: Human brain mapping: a systematic comparison of parcellation methods for the human cerebral cortex. Neuroimage **170**, 5–30 (2018)

6. Atluri, G., MacDonald III, A., Lim, K.O., Kumar, V.: The brain-network paradigm: using functional imaging data to study how the brain works. Computer **49**(10), 65–71 (2016)

7. Bearden, C.E., Thompson, P.M.: Emerging global initiatives in neurogenetics: the enhancing neuroimaging genetics through meta-analysis (ENIGMA) consortium. Neuron **94**(2), 232–236 (2017)

8. Beckmann, C.F., Mackay, C.E., Filippini, N., Smith, S.M.: Group comparison of resting-state FMRI data using multi-subject ICA and dual regression. Neuroimage **47**(Suppl 1), S148 (2009)

9. Blumensath, T., et al.: Spatially constrained hierarchical parcellation of the brain with resting-state fMRI. Neuroimage **76**, 313–324 (2013)

10. Braun, U., et al.: Test-retest reliability of resting-state connectivity network characteristics using fMRI and graph theoretical measures. NeuroImage **59**(2), 1404–1412 (2012)

11. Cao, H., et al.: Test-retest reliability of fMRI-based graph theoretical properties during working memory, emotion processing, and resting state. Neuroimage **84**, 888–900 (2014)

12. Craddock, R.C., James, G.A., Holtzheimer III, P.E., Hu, X.P., Mayberg, H.S.: A whole brain fMRI atlas generated via spatially constrained spectral clustering. Hum. Brain Mapp. **33**(8), 1914–1928 (2012)

13. Dubois, J., Adolphs, R.: Building a science of individual differences from fMRI. Trends Cogn. Sci. **20**(6), 425–443 (2016)

14. Eickhoff, S.B., Yeo, B.T., Genon, S.: Imaging-based parcellations of the human brain. Nat. Rev. Neurosci. **19**, 672–686 (2018)

15. Gordon, E.M., Laumann, T.O., Adeyemo, B., Huckins, J.F., Kelley, W.M., Petersen, S.E.: Generation and evaluation of a cortical area parcellation from resting-state correlations. Cereb. Cortex **26**(1), 288–303 (2014)

16. Jack Jr., C.R., et al.: The Alzheimer's disease neuroimaging initiative (ADNI): MRI methods. J. Magn. Reson. Imaging: Off. J. Int. Soc. Magn. Reson. Med. **27**(4), 685–691 (2008)

17. Jbabdi, S., Woolrich, M.W., Behrens, T.E.J.: Multiple-subjects connectivity-based parcellation using hierarchical Dirichlet process mixture models. NeuroImage **44**(2), 373–384 (2009)

18. Kelly, C., Biswal, B.B., Craddock, R.C., Castellanos, F.X., Milham, M.P.: Characterizing variation in the functional connectome: promise and pitfalls. Trends Cogn. Sci. **16**(3), 181–188 (2012)

19. Koo, T.K., Li, M.Y.: A guideline of selecting and reporting intraclass correlation coefficients for reliability research. J. Chiropr. Med. **15**(2), 155–163 (2016)

20. Meindl, T., et al.: Test-retest reproducibility of the default-mode network in healthy individuals. Hum. Brain Mapp. **31**(2), 237–246 (2010)

21. Murphy, K., Birn, R.M., Handwerker, D.A., Jones, T.B., Bandettini, P.A.: Intraclass correlations: uses in assessing rater reliability. Psychol. Bull. **86**(2), 420–428 (1979)
22. Murphy, K., Birn, R.M., Handwerker, D.A., Jones, T.B., Bandettini, P.A.: The impact of global signal regression on resting state correlations: are anti-correlated networks introduced? NeuroImage **44**(3), 893–905 (2009)
23. Potkin, S.G., Ford, J.M.: Widespread cortical dysfunction in schizophrenia: the FBIRN imaging consortium. Schizophr. Bull. **35**(1), 15–18 (2008)
24. Preti, M.G., Bolton, T.A., Van De Ville, D.: The dynamic functional connectome: state-of-the-art and perspectives. Neuroimage **160**, 41–54 (2017)
25. Rosen, B.R., Savoy, R.L.: fMRI at 20: has it changed the world? Neuroimage **62**(2), 1316–1324 (2012)
26. Thirion, B., Varoquaux, G., Dohmatob, E., Poline, J.B.: Which fMRI clustering gives good brain parcellations? Front. Neurosci. **8**, 167 (2014)
27. Thomas Yeo, B., et al.: The organization of the human cerebral cortex estimated by intrinsic functional connectivity. J. Neurophysiol. **106**(3), 1125–1165 (2011)
28. Tzourio-Mazoyer, N., et al.: Automated anatomical labeling of activations in SPM using a macroscopic anatomical parcellation of the MNI MRI single-subject brain. Neuroimage **15**(1), 273–289 (2002)
29. Van Essen, D.C., et al.: The WU-Minn human connectome project: an overview. Neuroimage **80**, 62–79 (2013)
30. Zilles, K., Amunts, K.: Centenary of Brodmann's map-conception and fate. Nat. Rev. Neurosci. **11**(2), 139 (2010)
31. Zuo, X.N., Kelly, C., Adelstein, J.S., Klein, D.F., Castellanos, F.X., Milham, M.P.: Reliable intrinsic connectivity networks: test-retest evaluation using ICA and dual regression approach. Neuroimage **49**(3), 2163–2177 (2010)

Constraining Disease Progression Models Using Subject Specific Connectivity Priors

Anvar Kurmukov[1,2(✉)], Yuji Zhao[3], Ayagoz Mussabaeva[1], and Boris Gutman[1,3]

[1] Institute for Information Transmission Problems, Moscow, Russia
kurmukovai@gmail.com
[2] National Research University Higher School of Economics, Moscow, Russia
[3] Department of Biomedical Engineering,
Illinois Institute of Technology, Chicago, IL, USA
bgutman1@iit.edu

Abstract. We propose a simple yet powerful extension for event-based progression disease model by exploiting the Network Diffusion Hypothesis. Our approach allows incorporating connectivity information derived from diffusion MRI data in the form of an informative prior on event ordering. This simple extension using a definition of transition probability based on network path length leads to improved reproducibility and discriminative power. We report experimental results on a subset of the Alzheimer's Disease Neuroimaging Initiative data set (ADNI 2). Though trained solely on cross-sectional data, our model successfully assigns higher progression scores to patients converting to more severe stages of dementia.

Keywords: Connectomes · Disease Progression Model · Alzheimer's Disease

1 Introduction

Imaging biomarkers of neurodegeneration have played an increasingly important role in clinical trials and disease stage assessment in recent years [1,2]. At the same time, as the trial design has grown increasingly complex, the very notion of "biomarker" has evolved. Classical notions of biomarker efficacy, such as the power under Normal assumptions [3,4], and classification accuracy [5] have given way to temporally aware models of disease [6,7]. This more recent approach to modeling disease, generally termed Disease Progression Modeling (DPM), assigns a time-dependent disease score (or stage) to each patient as a well as a canonical model of imaging and potentially non-imaging patient data as a function of this score. Unlike traditional classification approaches, this approach rests on the idea that different clinical and imaging features are discriminative

© Springer Nature Switzerland AG 2019
M. D. Schirmer et al. (Eds.): CNI 2019, LNCS 11848, pp. 106–116, 2019.
https://doi.org/10.1007/978-3-030-32391-2_11

at different stages of the disease: each marker has a specific finite time window during which it is affected by the illness. One of the earliest such models used with neuroimaging data is the Event-Based Model (EBM) [8]. Here, the disease score is treated as a discrete variable to be identified with a neurodegenerative "event" in each input phenotype, such as regional gray matter or the presence of misfolded proteins as measured with MRI or PET. The canonical event order is estimated by sampling from a Bayesian posterior formulation, generally using specific parametric distribution assumptions for healthy and diseased subjects. Variations on EBM include discriminative EBM [9], and simultaneous staging and unsupervised subject subtype identification [10]. Models beyond EBM allow for an explicit continuous-time reparameterization of each subject, effectively modeling both the continuous canonical form of all phenotypes in concert as well as individual "neurological reserve" of each patient. The fully longitudinal DPM's (LDPM), first proposed in [11] as a parametric sigmoidal progression function, were later expanded for spatially dense imaging features with additional spatial priors [12]. The parametric progression form was further relaxed in a Gaussian process formulation in [13].

A noteworthy aspect of the above methods is the lack of informative priors on the order in which specific phenotypes undergo degeneration. In fact, such a prior is readily discernible from available MRI data and has been used elsewhere. Specifically, Raj et al. proposed the Network Diffusion Hypothesis, whereby the neurodegenerative process develops in a highly stereotyped manner, according to the brain's structural connectivity [14]. One recent work on DPM has indeed fused these two ideas [15]. However, even there only a mean "standard" connectivity is used for all subjects. Here for the first time, we propose a subject-specific network prior to constrain DPM. We develop the idea in the context of EBM and apply the model to the ADNI 2 dataset. Initial results indicate a better longitudinal generalization compared to standard EBM, and better predictive ability of the resulting progression score when applied to new subjects.

The remaining paper is structured as follows: Sect. 2 describes the EBM model and introduces the connectivity prior. We explain our experimental pipeline in Sect. 3. Section 4 summarizes the results of our experiments. Finally, in Sect. 5 we discuss some possible enhancements of our approach and conclude.

2   Event-Based Models and the Connectivity Prior

2.1   The Event-Based Model

The idea of modeling disease progression in the form of distinct ordered events goes back to [8]. The authors propose that disease development could be divided into stages. Every stage is defined by an event, i.e. the moment when some biomarker switches its state from normal to abnormal. Specifically, given a set of M biomarkers $X = \{x_1, \ldots, x_M\}$ the EBM estimates order π (permutation of indexes $1 \ldots M$) in which each biomarker becomes abnormal. The model then takes the following form:

$$p(k|X,\pi) := \prod_{j=1}^{k} p(x_j|E_{\pi(j)}) \prod_{j=k+1}^{M} p(x_j|\neg E_{\pi(j)}). \tag{1}$$

This formula defines the probability $p(k|X,\pi)$ of being at progression stage k given a set of biomarkers X and an order π. Here, $p(x_j|E_{\pi(j)})$ and $p(x_j|\neg E_{\pi(j)})$ are likelihoods of a measurement x_j given that event $E_{\pi(j)}$ has or has not occurred. Given N subjects, the total likelihood of observing the data $\mathbf{X} = \{X_1, \ldots, X_N\}$ is the following:

$$p(\mathbf{X}|\pi) = \prod_{i}^{N} \sum_{k=0}^{M} p(k)p(k|X_i,\pi), \tag{2}$$

where $p(k)$ is the probability of being at stage k. $p(k)$ is typically treated as an uninformative (uniform) prior. Once (2) is maximized and an optimal π is found, one could easily evaluate disease stage k for every patient using the expression in (1). To find optimal π, one needs to maximize the posterior distribution, i.e. find the optimal order π given $\mathbf{X}$:

$$p(\pi|\mathbf{X}) \propto p(\pi)p(\mathbf{X}|\pi), \tag{3}$$

here $p(\pi)$ denotes the prior probability of specific order, which is also typically set to be uniform.

In the present paper, we suggest to use connectivity information obtained from diffusion MRI, to get an informative, non-uniform $p(\pi)$. This allows us to use personal information more directly since both $p(x|E)$ and $p(x|\neg E)$ are estimated on groups of subjects, but $p(\pi)$ could be computed for every subject separately. The original EBM uses longitudinal data but treats each observation as a separate. Here, we fit our model using exclusively cross-sectional data, which allows as to remove the possible effect of overfitting since observations from the same subject are highly correlated. Though our proposition could be easily implemented in any EBM extension, here we decided to use the original EBM, which allows us to isolate the effect of using the connectivity prior.

2.2   Connectome Prior via Path Probability

We now introduce some additional notation. As before we denote by $X = \{x_1, \ldots, x_M\}$ the set of biomarkers, in our case the gray matter thickness for M cortical regions. By G we denote connectivity matrix with exactly M nodes: $\{v_1, \ldots v_M\}$. Each node v_j uniquely corresponds to a biomarker x_j. We use the subscript i to denote different subjects, so X_i means subject i, x_{ij} means j-th biomarker of i-th subject and the same for G_i and v_{ij}. By $p(v_a \rightarrow v_b)$ we denote the probability of transitioning from node v_a to node v_b.

We define the probability of a specific path $\pi_{a,b,c} = \{v_a \rightarrow v_b \rightarrow v_c\}$ as a product of probabilities of every individual step:

$$p(\pi_{a,b,c}) := p(v_a \rightarrow v_b) \cdot p(v_b \rightarrow v_c). \tag{4}$$

next, we define the individual transition probability to be proportional to the shortest path between nodes:

$$p(v_a \rightarrow v_b) \propto e^{-\sigma(v_a, v_b)}, \qquad (5)$$

where $\sigma(v_a, v_b)$ denotes the shortest path between two nodes, thus transitioning to closer nodes is more probable. We normalize an exponent in such a way that all probabilities of transitioning from node v_a to all other nodes sum to 1, thus $p(v_a \rightarrow v_b) \neq p(v_b \rightarrow v_a)$. Specifically, for every connectome, we compute square $(M \times M)$ matrix of shortest paths S:

$$S_{a,b} = \sigma(v_a, v_b), \qquad (6)$$

next we apply an exponent as in Eq. (5), finally, we divide each element in each row by the sum over this row to make individual values sum to 1:

$$p(v_a \rightarrow v_b) = \frac{e^{-S_{a,b}}}{\sum\limits_{b=1}^{M} e^{-S_{a,b}}}. \qquad (7)$$

The intuition of Network Diffusion Hypothesis is the following: if some region becomes abnormal, it will affect other regions that are structurally closer (in terms of connectivity) faster than regions that are structurally farther. And the structural closeness of two nodes is the length of the shortest path between them.

2.3 Optimizing π

To find the optimal order of events π we need to optimize (3):

$$\pi^* = \arg\max_{\pi} p(\pi) p(\mathbf{X}|\pi)$$

$$= \arg\max_{\pi} p(\pi) \prod_{i}^{N} \sum_{k=0}^{M} p(k|X_i, \pi) = \arg\max_{\pi} \prod_{i}^{N} p(\pi)^{1/N} \sum_{k=0}^{M} p(k|X_i, \pi)$$

$$= \arg\max_{\pi} \sum_{i}^{N} \log \left[p(\pi)^{1/N} \sum_{k=0}^{M} p(k|X_i, \pi) \right]$$

$$= \arg\max_{\pi} \frac{1}{N} \sum_{i}^{N} \log p(\pi) + \sum_{i}^{N} \log \left[\sum_{k=0}^{M} p(k|X_i, \pi) \right], \qquad (8)$$

One could compute $p(\pi)$ based on the average connectome $(\hat{(G)} = \frac{1}{N} \sum_{i}^{N} G_i)$. However, we find that using $\sum_{i}^{N} \log p_i(\pi)$ instead of $\sum_{i}^{N} \log p(\pi)$ leads to much better results. In other words, every $p_i(\pi)$ is computed from the corresponding individual connectome; for every subject, the prior on the order of events π is different.

As the Bayesian formulation with the connectome prior is identical in form, the optimization of (8) can be done using the MCMC procedure exactly as in [8].

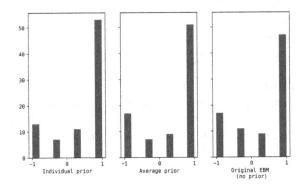

Fig. 1. Agreement results. Agreement between stages and visit order was measured using Kendall tau. Under the assumption that abnormality is irreversible, we measure Kendall tau between the vector of subject visits order and vector of subject disease progression scores. This figure summarizes the distribution of Kendall tau over all subjects.

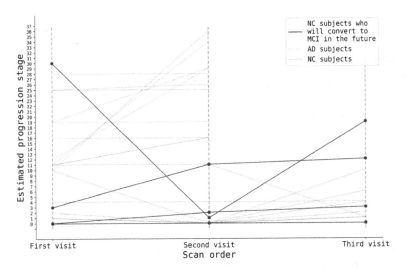

Fig. 2. Selected ADNI participants. Progression scores using order obtained with individual prior. Progression scores were computed using formula (1). Once we obtain optimal order we compute progression score for every subject at each time point.

3   Experiments

3.1   Data

Our data consisted of 84 unique subjects from the ADNI 2 dataset: 32 Alzheimer's patients, including 8 women (mean age 76.8 ± 7.6) and 24 men (mean age 75.9 ± 7.8) (AD), and 52 cognitively normal controls, including 26 women (mean age 72.0 ± 4.9) and 26 men (mean age 73.6 ± 6.4) (NC). We used

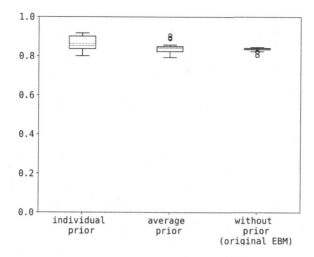

Fig. 3. Binary classification results. Classification was done based on subject stage. Performance was measured in terms of ROC AUC. Classification uncertainty was measured using 200 independent MCMC runs. All models were trained on cross-sectional data and performance was measured on longitudinal data (excluding training observations).

anatomical MRI data from 3 visits for each NC subject, and 2 visits for AD subjects. Of the control subjects, 4 are known future converters to MCI. Regional gray matter thickness was obtained using FreeSurfer 5.3, based on the Dessikan-Killiany atlas. We used only the baseline visit diffusion MRI to construct individual connectomes. Briefly, we used FSL's eddie correction and ANTs SyN for EPI artifact correction to T1 MRI. To extract streamlines, we used constrained spherical deconvolution (CSD) with a probabilistic tractography algorithm, as implemented in Dipy [16]. Finally, weighted connectivity matrices G have 0 on a main diagonal and the weights of edges are inversely proportional to the logarithm of the number of streamlines:

$$G_{a,b} = \begin{cases} \frac{1}{\log(1+w_{a,b})}, \text{if } w_{a,b} > 0 \\ 0, \text{if } w_{a,b} = 0 \end{cases} \qquad (9)$$

where $G_{a,b}$ is the edge between nodes a and b; $w_{a,b}$ is the number of streamlines between corresponding regions. The idea behind this specific weighting scheme is the following: firstly, we need the edges to be inversely proportional to number of streamlines (so the weight on edges has notion of distance not similarity); secondly, we do not want to penalize weak connections too much, so we take logarithm; finally, for streamlines with weight 1 we want an edge in a resulting connectome, so we add 1.

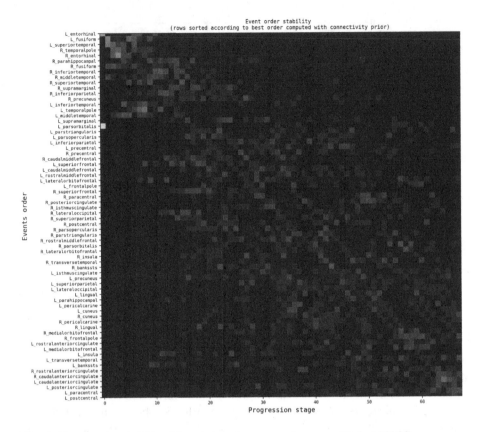

Fig. 4. Event order stability. Uncertainty is measured over multiple MCMC runs, rows are sorted according to best order (using individual prior)

3.2 Experimental Pipeline

We compare three different versions of the EBM:

1. The original EBM [8].
2. EBM with connectivity prior obtained from average connectome.
3. EBM with individual connectivity priors.

All models were trained on a first time point (cross-sectional data) observation and tested on latter time points (longitudinal data). Recall that evaluation of subject stage using formula (1) does not require $p(\pi)$, but only subject feature vector $X_i = \{x_{i1}, \ldots, x_{iM}\}$ and optimal π^*. Positional uncertainty of the orders was measured over multiple (200) MCMC runs. As additional performance markers, we used ROC AUC in using the inferred disease stage to discriminate AD and NC subjects.

4   Results

The natural way to measure agreement between the predicted stage of a disease based on anatomical features and each subject's actual visit order is with an ordinal correlation. Kendall's τ is the simplest choice, which we use here. We display the distribution of τ over all subjects in Fig. 1. Mean τ for standard EBM was 0.34, for mean connectome prior 0.41, and for individual connectome prior 0.49. Predicted disease stage for the subjects, including the 4 converters, is displayed in Fig. 2.

Classification accuracy followed a similar progression, improving with the mean connectome prior, and improving further with the individual prior (Fig. 3). Mean (standard deviation) ROC AUC over 200 independent MCMC optimizations was 0.816 (0.008) for standard EBM, 0.83 (0.026) for EBM with mean connectome prior, and 0.88 (0.046) for EBM with individual connectome prior.

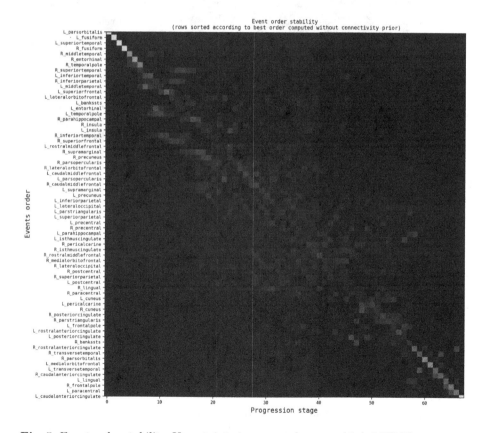

Fig. 5. Event order stability. Uncertainty is measured over multiple MCMC runs, rows are sorted according to best order (no prior)

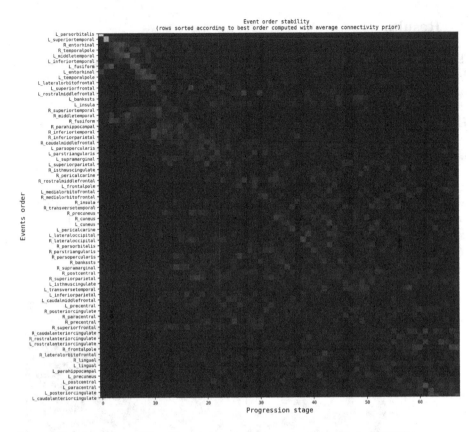

Fig. 6. Event order stability. Uncertainty is measured over multiple MCMC runs, rows are sorted according to best order (using average prior)

Finally, in Figs. 4, 5 and 6 we display the region by order probability matrix as an indicator of the stability of the canonical order computation. Unsurprisingly, using additional individual prior information makes the canonical order significantly less stable. This suggests that the overall EBM model with a single canonical order may not be sufficient to capture true subject variability in the way disease affects different brain regions over time.

Code reproducing all the results is published online.[1]

5   Conclusion

We have presented a direct way to incorporate the Network Diffusion Hypothesis into an established disease progression model. The extension via an informative prior improves several aspects of biomarker performance, including classification accuracy, and conversion prediction. Importantly, the work highlight the need to

---

[1] https://github.com/kurmukovai/ebm-connectivity-prior.

develop more sophisticated models of disease progression that take into account individual differences in brain connectivity and the resulting manner in which the disease and specific symptoms are likely to progress. This may include updating the longitudinal DPM models, for example by placing priors on sigmoidal progression parameters, as well as entirely new formulations that replace the notion of a canonical progression with a two-level stochastic process. Implications of the improved progression modeling include better subject stratification for clinical trials, lower drug development costs, and more accurate prediction of future cognitive decline.

Acknowledgments. Work by BG was supported by the Alzheimer's Association grant 2018-AARG-592081, Advanced Disconnectome Markers of Alzheimer's Disease. AK and AM were supported by the Russian Science Foundation under grant 17-11-01390.

References

1. Márquez, F., Yassa, M.A.: Neuroimaging biomarkers for Alzheimer's disease. Mol. Neurodegener. **14**(1), 21 (2019)
2. Petrella, J.R., Hao, W., Rao, A., Murali Doraiswamy, P., Alzheimer's Disease Computational Modeling Initiative: Computational causal modeling of the dynamic biomarker cascade in Alzheimer's disease (2019)
3. Beckett, L.A.: Community-based studies of Alzheimer's disease: statistical challenges in design and analysis. Stat. Med. **19**, 1469–1480 (2000)
4. Gutman, B.A., et al.: Empowering imaging biomarkers of Alzheimer's disease. Neurobiol. Aging **36**(Suppl 1), S69–S80 (2015)
5. Gao, C., et al.: Model-based and model-free machine learning techniques for diagnostic prediction and classification of clinical outcomes in Parkinson's disease. Sci. Rep. **8**(1), 7129 (2018)
6. Jack Jr., C.R., et al.: Hypothetical model of dynamic biomarkers of the Alzheimer's pathological cascade. Lancet Neurol. **9**(1), 119–128 (2010)
7. Oxtoby, N.P., Alexander, D.C.: Imaging plus X: multimodal models of neurodegenerative disease. Curr. Opin. Neurol. **30**(4), 371 (2017)
8. Fonteijn, H.M., et al.: An event-based model for disease progression and its application in familial Alzheimer's disease and Huntington's disease. NeuroImage **60**(3), 1880–1889 (2012)
9. Venkatraghavan, V., et al.: Disease progression timeline estimation for Alzheimer's disease using discriminative event based modeling. NeuroImage **186**, 518–532 (2019)
10. Young, A.L., et al.: Multiple orderings of events in disease progression. In: Ourselin, S., Alexander, D.C., Westin, C.-F., Cardoso, M.J. (eds.) IPMI 2015. LNCS, vol. 9123, pp. 711–722. Springer, Cham (2015). https://doi.org/10.1007/978-3-319-19992-4_56
11. Jedynak, B.M., et al.: A computational neurodegenerative disease progression score: method and results with the Alzheimer's disease neuroimaging initiative cohort. Neuroimage **63**(3), 1478–1486 (2012)
12. Marinescu, R.V., et al.: DIVE: a spatiotemporal progression model of brain pathology in neurodegenerative disorders. Neuroimage **192**, 166–177 (2019)

13. Lorenzi, M., Filippone, M., Frisoni, G.B., Alexander, D.C., Ourselin, S.: Probabilistic disease progression modeling to characterize diagnostic uncertainty: application to staging and prediction in Alzheimer's disease. NeuroImage **190**, 56–68 (2019)
14. Raj, A., Kuceyeski, A., Weiner, M.: A network diffusion model of disease progression in dementia. Neuron **73**(6), 1204–1215 (2012)
15. Garbarino, S., Lorenzi, M.: Modeling and inference of spatio-temporal protein dynamics across brain networks. In: Chung, A.C.S., Gee, J.C., Yushkevich, P.A., Bao, S. (eds.) IPMI 2019. LNCS, vol. 11492, pp. 57–69. Springer, Cham (2019). https://doi.org/10.1007/978-3-030-20351-1_5
16. Garyfallidis, E., et al.: Dipy, a library for the analysis of diffusion MRI data. Front. Neuroinform. **8**, 8 (2014)

Hemodynamic Matrix Factorization for Functional Magnetic Resonance Imaging

Michael Hütel[1]($\boxtimes$), Michela Antonelli[1], Jinendra Ekanayake[1,2,3],
Sebastien Ourselin[1], and Andrew Melbourne[1]

[1] School of Biomedical Engineering and Imaging Sciences, KCL, London, UK
michael.hutel@kcl.ac.uk
[2] Wellcome/EPSRC Centre for Interventional and Surgical Sciences,
UCL, London, UK
[3] Department of Neurosurgery, Leeds General Infirmary, Leeds, UK

Abstract. Neural activation causes a complex change in neurophysiological parameters of the cerebral blood flow (CBF). Functional magnetic resonance imaging (fMRI) measures one of these neurophysiological parameters, which is the blood oxygen level dependent (BOLD) response. The general linear model (GLM) used in fMRI task experiments relates activated brain areas to extrinsic task stimuli. The translation of task-induced neural activation into a hemodynamic response is approximated with a convolution model in the GLM design. There are major limitations to the GLM approach. First, the GLM approach does not model intrinsic brain activity. Second, the GLM assumes compliant task participation matching the stimulus timing and duration in the corresponding task. We propose hemodynamic matrix factorization (HMF), a data-driven approach to model intrinsic and extrinsic neural activation in fMRI. By contrast to the GLM, the HMF does not incorporate the original task design. The neural activation is a latent variable and estimated from fMRI data. Each component of the HMF consists of a neural activation time course and a spatial mapping. A linear filter translates neural activation time courses into BOLD responses. We apply our HMF to a motor localization task of an open source data cohort. We obtain neural activation time courses that correlate with the original block design of the task and whose corresponding spatial maps match individual areas of the sensory-motor cortex known to be activated by either foot, hand or tongue movement. We find HMF components whose neural activation time courses correlate with the visual cue timings presented at the beginning of each task block. HMF thus constitutes a novel tool to validate if the actual task execution of a subject matches the intended execution specified in the task design of fMRI experiments.

Electronic supplementary material The online version of this chapter (https://doi.org/10.1007/978-3-030-32391-2_12) contains supplementary material, which is available to authorized users.

1   Introduction

One of the most defining assumptions in task experiments is that neural activity is an idealized variable that is either on or off following the exact timing of a presented stimulus. This 'idealized neural activity' assumption has shaped the task experiment literature for decades and constitutes the origin of hypothesis-driven models in fMRI. The most prominent hypothesis-driven model is the GLM [1]. The design of the GLM is based on the following assumptions: (a) the stimulus presentation during a task experiment results in an immediate increase in neural activation; (b) the increase in neural activity translates into a BOLD signal change by a filter operation following the laws of a linear time-invariant (LTI) system [2]. The impulse response function of this LTI system is referred to as hemodynamic response function (HRF) in fMRI. The estimates of the GLM result in a statistical parametric mapping (SPM) showing to what degree individual areas of the brain are involved in the stimulus processing.

In contrast to the hypothesis-driven GLM, there are data-driven methods such as spatial independent component analysis (ICA) [3] that factorizes fMRI data into a set of timecourses and spatial maps by optimizing a proxy function for spatial independence.

We propose HMF, a novel data-driven technique that decomposes fMRI data into a set of components, where each component consists of a latent neural activation time course and a spatial map. The neural activation time course is translated into a BOLD time course with the canonical HRF. Blind to the original stimulus timing, HMF recovers neural activation time courses that match the idealized neural activation assumed in the motor task experiment and reveal non-compliant subject behavior.

2   Hemodynamic Matrix Factorization

The proposed generative model is concerned with BOLD time courses of length T for S subjects, measured on V voxels: $\{\mathbf{Y_s} \in \mathbb{R}^{T \times V}, s = 1 \cdots S\}$. We assume that the signal change observed in all time-concatenated subjects $\mathbf{Y} \in \mathbb{R}^{(T*S) \times V}$ is driven by C latent components that form depending on the performed task. Our model assumes that each component consists of a BOLD time course $b_c \in \mathbb{R}^{T \times 1}$ and a spatial map $h_c \in \mathbb{R}^{1 \times V}$. The BOLD time course $b_c = n_c \circledast f$ is obtained by convolving a neural activation time course n_c with a canonical haemodynamic response function (HRF) f. Given that multiple time courses are convolved with the same filter f, the convolution operation can be rewritten as matrix multiplication of a time course matrix $\mathbf{N} \in \mathbb{R}^{T \times C}$ with Töplitz matrix $\mathbf{F} \in \mathbb{R}^{T \times T}$ of filter f such that $\mathbf{B} = \mathbf{FN}$. The proposed generative model for the data $\mathbf{Y}$ is then obtained by C components such that

$$\mathbf{Y} = \mathbf{FNH}, \tag{1}$$

where matrix $\mathbf{N} \in \mathbb{R}^{T*S \times C}$ contains the individual neural activation time courses per subject $\mathbf{N_s} \in \mathbb{R}^{T \times C}$ and matrix $\mathbf{H} \in \mathbb{R}_+^{C \times V}$ contains a set of spatial maps shared among all subjects.

We minimize the error between the actual data $\mathbf{Y}$ and the approximated data generated by the model $\mathbf{FNH}$, resulting in the following cost function:

$$J = \frac{1}{2}\|\mathbf{M} \odot (\mathbf{FNH} - \mathbf{Y})\|_2^2 = \frac{1}{2}\|\mathbf{M} \odot (\mathbf{FN}\sigma((\mathbf{FN})^\top\mathbf{Y} + b) - \mathbf{Y})\|_2^2, \quad (2)$$

where the matrix $\mathbf{M} \in \mathbb{R}^{1 \times V}$ weighs the importance of each individual voxels.

The regularization for the spatial map matrix $\mathbf{H}$ and neural activation matrix $\mathbf{N}$ are introduced in the following.

Regularization of Spatial Maps $\mathbf{H}$. To enforce sparse spatial maps, we use the Kullback Leibler divergence between two exponential distributions $\Delta(p\|q(H_c)) = \log(\lambda) - \log(\hat{\lambda}_c) + \frac{\hat{\lambda}_c}{\lambda} - 1$ between desired rate parameter λ and estimated rate parameter $\hat{\lambda}_c$ of an exponential value distribution approximating the value distribution in each spatial map $H_c \in \mathbb{R}^{1 \times V}$, using the following regularization term:

$$R_{IS} = \frac{1}{C}\sum_{c=1}^{C}\Delta(p\|q(H_c)). \quad (3)$$

In addition to sparsity, we want to enforce smoothness of the spatial maps using anisotropic total variation. Given that we can reshape spatial map matrix $\mathbf{H}$ in a 4D tensor $\mathbf{H}^{(4D)}$, we want to minimize the following regularization cost:

$$\begin{aligned} R_{\mathbf{H}} = \sum_{c,i,j,k} &\left|\mathbf{H}^{(4D)}{}_{c,i+1,j,k} - \mathbf{H}^{(4D)}{}_{c,i,j,k}\right| \\ &+ \left|\mathbf{H}^{(4D)}{}_{c,i,j+1,k} - \mathbf{H}^{(4D)}{}_{c,i,j,k}\right| + \left|\mathbf{H}^{(4D)}{}_{c,i,j,k+1} - \mathbf{H}^{(4D)}{}_{c,i,j,k}\right| \end{aligned} \quad (4)$$

The product of $\mathbf{H} = \sigma((\mathbf{FN})^\top\mathbf{Y} + b)$ can be rewritten element-wise as

$$\left[\sigma((\mathbf{FN})^\top y_1 + b)\ \sigma((\mathbf{FN})^\top y_2 + b)\ \dots\ \sigma((\mathbf{FN})^\top y_V + b)\right], \quad (5)$$

where $\sigma((\mathbf{FN})^\top y_v + b)$ constitutes a column vector for voxel v. All column vectors taken together form the spatial maps $\mathbf{H}$ as depicted in Eq. 5. Furthermore, one can rearrange the original data $\mathbf{Y}$ in individual matrices along either dimension $V1$, $V2$ or $V3$ of the three-dimensional Euclidean space. An approximation of Eq. 4 can then be computed with the following term

$$\begin{aligned} R_{\mathbf{H}} = &\sum_{i}^{V2}\sum_{j}^{V3}\|\sigma((\mathbf{FN})^\top\mathbf{Y}_{ij}^{(1)} + b)\mathbf{D}_1^\top\|_1 \\ &+ \sum_{i}^{V1}\sum_{j}^{V3}\|\sigma((\mathbf{FN})^\top\mathbf{Y}_{ij}^{(2)} + b)\mathbf{D}_2^\top\|_1 + \sum_{i}^{V1}\sum_{j}^{V2}\|\sigma((\mathbf{FN})^\top\mathbf{Y}_{ij}^{(3)} + b)\mathbf{D}_3^\top\|_1 \end{aligned}$$

$$(6)$$

where $\mathbf{Y}_{ij}^{(1)} \in \mathbb{R}^{T \times V1}$, $\mathbf{Y}_{ij}^{(2)} \in \mathbb{R}^{T \times V2}$, $\mathbf{Y}_{ij}^{(3)} \in \mathbb{R}^{T \times V3}$, $\mathbf{D_1} \in \mathbb{R}^{(V1-1) \times V1}$, $\mathbf{D_2} \in \mathbb{R}^{(V2-1) \times V2}$ and $\mathbf{D_3} \in \mathbb{R}^{(V3-1) \times V3}$. The difference operator matrices $\mathbf{D_i}$ have the form

$$
\mathbf{D_i} = \begin{pmatrix} -1 & 1 & & & & \\ & -1 & 1 & & & \\ & & -1 & 1 & & \\ & & & \ddots & & \\ & & & & -1 & 1 \\ & & & & & -1 & 1 \end{pmatrix} \tag{7}
$$

Regularization of Neural Activation $\mathbf{N}$. The approximate anisotropic total variation proposed for spatial regularization in Eq. 7 can also be applied as temporal regularization to the neural time course matrix $\mathbf{N}$ with corresponding cost function

$$
R_{\mathbf{N}} = \|\mathbf{N}\mathbf{D_4}\|_1, \tag{8}
$$

where $\mathbf{D_4} \in \mathbb{R}^{(T*S-1) \times (T*S)}$.

The overall cost $\Omega = J + R_N + R_{IS} + R_H$ is optimized with respect to neural activation time courses $\mathbf{N}$.

3   Experiments and Results

3.1   Experimental Setup

We applied HMF to the motor task of the Midnight scan club (MSC) [4] open source data[1], which comprises 10 healthy adult subjects performing the same motor task in 10 different sessions.

The motor task design consists of 5 repeated blocks, in which a subject performs one of the 5 movement types, which are left or right foot, left or right hand, or tongue movement. There is a visual cue at the beginning of each block instructing the subject of what movement to conduct. For each fMRI scan, volumes were aligned to the first volume to correct for head motion. The first volume was linearly registered to its corresponding bias-corrected T1-weighted anatomical scan. The intra-subject affine registration and non-linear registration to the Montreal Neurological Institute (MNI) template were combined to map all fMRI volumes with one re-sampling into the MNI space sampled in a 4mm isotropic resolution. Time courses of voxels within brain tissue were extracted, high-pass filtered (0.01Hz cut-off) to remove signal drifts from scanner instabilities, centered and variance-normalized. The functional 4D volume set per scan s was reshaped into a matrix $\mathbf{Y}_s \in \mathbb{R}^{T \times V}$ with T time points and V voxels.

In the following section, we compare the neural activation time courses of the HMF components to the stimulus timing of each task. We then compare the corresponding BOLD time courses of these HMF components to the timecourses obtained by spatial ICA. ICA decomposes the data $\mathbf{Y} = \mathbf{B}\mathbf{H}$ into a matrix of

---

[1] https://openfmri.org/dataset/ds000224/.

BOLD time courses $\mathbf{B} \in \mathbb{R}^{T*S \times C}$ and a matrix of spatial maps $\mathbf{H} \in \mathbb{R}^{C \times V}$. For both decompositions, we set the number of components C to either 20, 40, 60, 80, or 100 components.

3.2 Neural Activation

Figure 1 shows the spatial map and neural activation time course of the 5 most correlated HMF components to the visual cues (decreasing top to bottom) when $C = 100$. The spatial maps obtained by HMF match with anatomical areas that are likely involved in the motor task processing. The spatial maps shown in the first and fourth row of Fig. 1, are related to visual stimulus processing. Whereas the second, third and fifth row show spatial maps that comprise anatomical areas involved in motor planning and execution, such as supplementary motor cortex and basal ganglia network.

Figure 2 depicts the five individual block timings (dotted lines) related to the visual cue shown in Fig. 1. Each row corresponds to one of the 5 movements: left foot (first row), right foot (second row), left hand (third row), right hand (fourth row) and tongue (fifth row). The HMF decomposition provides a good model for hand and tongue movements as the estimated neural activation time course closely resembles the idealized neural activation assumed for the task. It is evident from Fig. 2 that the noise level in subject 4, results in a poor model fit, likely due to head motion.

Most importantly, HMF is able to detect that subject 6 performed a right hand movement instead of a left hand movement and vice versa in the second task block (red box in Fig. 2). Finally, the neural activation time courses for the foot movement not always resemble the expected idealized neural activation for all subjects. Indeed, the neural activation time course in subject 1 closely resembles the task design whereas there is overlap between neural activation in left and right foot HMF component for subject 2.

3.3 Comparison of HMF and ICA

We compare BOLD time course and spatial map of the HMF components to the respective equivalents of ICA. Figure 3 depicts the distribution of correlation values of the most correlated HMF component to each individual movement and visual timing convolved with the canonical HRF (i.e., left and right toe, left and right hand and tongue). Figure 4 shows the reproducibility of the spatial map of the one component associated with an individual movement or visual cue in each session. The reproducibility is defined by the pair-wise spatial correlation between corresponding components of two distinct sessions.

HMF and ICA produced components with similar spatial and temporal structure as presented in the supplementary material in Figs. 6 and 7, respectively. This is evident in Fig. 3 where the correlation between the BOLD time courses and the task design increases with the increase of the number of components C. HMF components presented with a higher median correlation than the components obtained by ICA for all the task and all values of C except for the

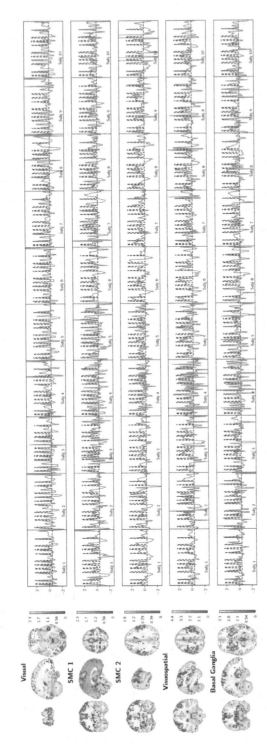

Fig. 1. Spatial map and neural activation time course (continuous line) of the 5 most correlated HMF components to the visual cue timing (dotted line) for the first session of each subject. (SMC = Supplementary Motor Cortex)

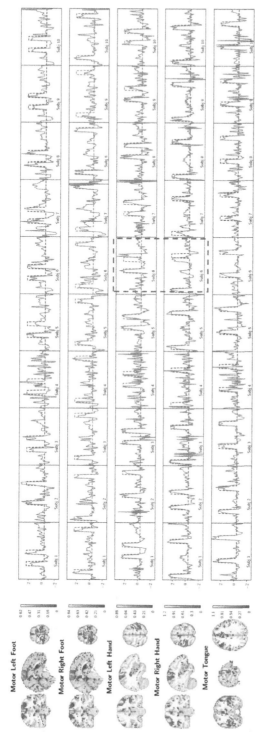

Fig. 2. Spatial map and neural activation time course (continuous line) of the 5 different movements, namely, left and right hand, left and right foot, and tongue (from the top to bottom), of the most correlated HMF component to the block timing of the movement (dotted line) for the first session of each subject. The red dotted box shows subject 6 confusing right and left hand movement in the second task block. (Color figure online)

left and right foot when $C = 80$. Moreover, the spatial maps of HMF are more reproducible than the spatial maps of ICA as shown in Fig. 3.

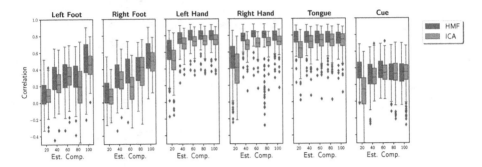

Fig. 3. The distribution of temporal correlation of HMF and ICA components with the corresponding task timing.

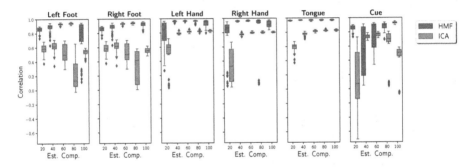

Fig. 4. The distribution of spatial correlation between components of different task sessions that relate to the same task timing.

4   Discussion

We proposed a novel data-driven technique called HMF that decomposes fMRI data into components whose neural activation time courses resemble timings of the respective visual stimulation and movement execution timing. Most importantly, the timing information of individual stimuli and task execution were not known to HMF. For the visual stimulation, HMF identified 5 components that were highly correlated with visual cue timings suggesting their involvement in the processing of the visual cues as well as the contemplation of subsequent motor execution in the movement blocks. Regarding the movement execution, HMF identifies only one highly correlated component with the block timings (see Fig. 5 in supplementary material) for each type of the 5 movements.

We compared HMF to ICA, which is the most common data-driven tool to infer brain networks from spontaneous intrinsic neural activity. The results showed that both techniques produced task-related components whose BOLD time courses and spatial maps are highly similar. However, HMF produced on average more correlated components than ICA. This is likely due to the regularization of HMF that relates adjacent points in time and space in fMRI data, whereas there is no such regularization in ICA.

In contrast to data-driven techniques, hypothesis-driven techniques such as GLM require assumptions of how the task design relates to neural activation and subsequent change in BOLD. Although HMF did not use such task timing information, it generated spatial maps that matched the spatial activation maps presented in [4]. The GLM provides a statistical test framework to infer brain activation maps involved in the task processing. However, in contrast to data-driven techniques, the GLM cannot model intrinsic brain activity and relies crucially on task participation to obtain correct statistical test estimates.

If task participation is different than expected, statistical inference within the GLM framework will lead to a wrong statistical estimates. Tools such as HMF are therefore needed to validate task participation, especially in particular cohorts such as children that are likely to be less task compliant.

Data-driven techniques produce varying results depending on hyper-parameters, i.e. the number of components or the strength of applied regularization. Inferring neural activation from BOLD is an ill-posed problem given that multiple neural activation patterns can produce the exact same BOLD time course. We apply total variation regularization assuming that the most plausible is likely the most variational simple solution. However, sensible hyper-parameters can be found by finding a balance between reproducibility and explained variance.

We have developed a robust physiology-based methodology to analyze task-based fMRI data at the individual level. Methods such as this will help us to guide our future understanding of the processes of cognition.

References

1. Poline, J.B., Brett, M.: The general linear model and fMRI: does love last forever? Neuroimage **62**(2), 871–880 (2012)
2. Boynton, G.M., Engel, S.A., Heeger, D.J.: Linear systems analysis of the fMRI signal. NeuroImage **62**(2), 975–984 (2012)
3. Beckmann, C.F., Smith, S.M.: Probabilistic independent component analysis for functional magnetic resonance imaging. IEEE Trans. Med. Imaging **23**(2), 137–152 (2004)
4. Gordon, E.M., et al.: Precision functional mapping of individual human brains. Neuron **95**(4), 791–807 (2017)

Network Dependency Index Stratified Subnetwork Analysis of Functional Connectomes: An Application to Autism

Ai Wern Chung[1] and Markus D. Schirmer[2,3]

[1] Fetal-Neonatal Neuroimaging and Developmental Science Center,
Boston Children's Hospital, Harvard Medical School, Boston, MA, USA
[2] Stroke Division & Massachusetts General Hospital,
J. Philip Kistler Stroke Research Center, Harvard Medical School, Boston, USA
mschirmer1@mgh.harvard.edu
[3] Department of Population Health Sciences,
German Centre for Neurodegenerative Diseases (DZNE), Bonn, Germany

Abstract. Autism spectrum disorder (ASD) is a neurodevelopmental condition impacting high-level cognitive processing and social behavior. Recognizing the distributed nature of brain function, neuroscientists are exploiting the connectome to aid with the characterization of this complex disease. The human connectome has demonstrated the brain to be a highly organized system with a centralized core vital for effective function. As such, many have used this topological principle to not only assess core regions, but have stratified the remaining graph into subnetworks depending on their relation to the core. Subnetworks are then utilized to further understand the supporting role of more peripheral nodes with respects to the overall function in the network. A recently proposed framework for subnetwork definition is based on the network dependency index (NDI), a measure of a node's importance based on its contribution to overall efficiency in the network, and the derived subnetworks, or Tiers, have been shown to be largely stable across ages in structural networks. Here, we extend the NDI framework to test its efficacy against a number experimental conditions. We first not only demonstrated NDI's feasibility on resting-state functional MRI data, but also its stability irrespective of the group connectome on which NDI was determined for various edge thresholds. Secondly, by comparing network theory measures of transitivity and efficiency, significant group differences were identified in NDI Tiers of greatest importance. This demonstrates the efficacy of utilizing NDI stratified subnetworks, which can help to improve our understanding of diseases and how they affect overall brain connectivity.

Keywords: Network dependency index · Subnetworks · Autism · Functional · Connectome · rsfMRI

© Springer Nature Switzerland AG 2019
M. D. Schirmer et al. (Eds.): CNI 2019, LNCS 11848, pp. 126–137, 2019.
https://doi.org/10.1007/978-3-030-32391-2_13

1   Introduction

Neurodevelopmental conditions impair the growth and/or development of the brain. One such widely studied condition is autism spectrum disorder (ASD), which affects about 1.7% of children in the US [1]. ASD describes a spectrum of neurodevelopmental disorders characterized by atypical social behavior and sensory processing, where patients also demonstrate deficits in mental flexibility and high-level cognitive function [11,16], and is currently diagnosed using cognitive assessment. Investigations suggest that ASD is a distributed disease and cannot be described by local effects, i.e. by specific brain regions, leading to an increased interest in applying connectomics for identifying differences in the autistic brain [11,13].

Brain connectivity, or connectomics, and its topology has been widely studied, e.g., in healthy subjects [5,10,18,21,24], during early brain development [2,6,17], and in disease [6,7,12,20]. In particular, studies have used various ways to define subnetworks in the human connectome. These studies stratify groups of nodes in a brain network by a network theoretical measure, often relating it to the underlying network topology. Subsequent analyses often compare "traditional" network measures between groups within the cohort or between subnetworks [4,7,18,19]. However, most subnetwork stratification, after the brain network has been estimated, relies on a user-defined parameter, which can have significant impact on the subnetwork definition (e.g. k in rich-club analyses [21]). A recent study investigated the use of the network dependency index (NDI) to identify subnetworks in a data driven fashion and subsequently no user-parameter needs to be defined. In their study, Schirmer et al. [19] utilized structural connectomes with 170 brain regions in the NKI-Lifespan cohort to investigate the stability of NDI across age groups, and compared their method with subnetworks defined using the rich-club. NDI assigns a measure of "importance" to each node in the connectome by quantifying the global effect of removing the node on network efficiency. Subsequent subnetwork stratification groups nodes automatically into Tiers based on this measure, identifying sets of nodes which can be considered essential for network efficiency. Importantly, their NDI framework demonstrated high reliability in determining a consistent set of regions to belong to the same subnetworks, without having to specify a user-defined parameter. However, they did not investigate the feasibility of applying their framework to functional data, or the utility of the identified subnetworks to identify group differences in a patient-control setting.

In this work, we apply the NDI framework to a set of resting-state functional connectomes in an ASD/control cohort based on the Automated Anatomical Labeling (AAL) atlas. We demonstrate that the framework can directly be applied to functional connectomes that are parcellated by a commonly used atlas on which few regions are defined. We also investigate the effect of using the cohort, control-only, and patient-only connectomes to derive NDI subnetworks. Additionally, we investigate the consistency of nodal subnetwork assignment by using different weighting schemes in the functional connectome, i.e. retaining only edges with negative weights, positive weights, and lastly the absolute weights of both, while varying the threshold for noise removal. Finally, we compare topological features in each of the subnetworks generated by all weighting schemes between subjects diagnosed with ASD and typically developing individuals.

2   Materials and Methods

2.1   Study Design and Patient Population

Data used in this study originates from the Autism Brain Imaging Data Exchange (ABIDE) [8,9] initiative and was downloaded through the python package *nilearn* [14]. ABIDE consists of data comprising ASD (patients) and typically developing (controls) individuals [9]. Each individual underwent a magnetic resonance imaging protocol, including rsfMRI and MPRAGE sequences. Details of acquisition, informed consent, and site-specific protocols are available elsewhere[1]. The cohort characteristics are summarized in Table 1.

Table 1. Cohort characterization.

	Cohort	Control	Patients
N	819	440	379
Age, years, mean (sd)	16.39 (7.12)	16.27 (6.74)	16.53 (7.54)

2.2   RsfMRI Preprocessing and Group Connectomes

Data were preprocessed based on the ABIDE Connectome Computation System pipeline, which included slice timing and motion correction, removal of mean CSF and white matter signals, and detrending of linear and quadratic drifts. Subsequently, a temporal band-pass filtering was applied (0.01–0.1 Hz) and the rsfMRI data registered to the MNI template. Regions for network analysis were defined based on the AAL atlas and the pre-processed time series was demeaned. Prior to network analysis, brainstem and cerebellar regions were removed, resulting in a total of 90 brain regions. Edge weights were computed as a covariance matrix [22] and edges with an absolute weight less than a given threshold were removed to reduce the effects of spurious signals. In this study, we investigate edge weight thresholds of 0.01, 0.03, and 0.05. In this study, we investigate three kinds of networks - by retaining only the positive weights (pos), only the absolute values of negative weights (neg), and the absolute of all weights (abs) - as there is no consensus on which of these are most discriminative.

Subnetworks may be determined on group-averaged connectomes. Such connectomes have been used in multiple studies [18,19,21]. First, the binarized

---

[1] http://fcon_1000.projects.nitrc.org/indi/abide/.

connectivity matrices of all subjects within a group are summarized by only retaining edges that are present in at least 90% of the subjects (group adjacency matrix) with the goal to preserve connections which can be reliably identified. To allow for weighted network analyses, weights are added to the edges of the group adjacency matrix by averaging the edge weights from the contributing subjects' connectivity matrices. This process creates a weighted group-averaged connectome W_{group}, which can be utilized for analyses.

In this study, we group our cohort in three different ways. First, we create a *cohort connectome*, which utilizes information from all subjects within the cohort. As our cohort contains patients and controls, we also generate a *patient connectome*, as well as a *control connectome* for NDI analysis.

2.3 Network Dependency Index Subnetworks

A detailed description of the NDI framework is given elsewhere [19]. In brief, given a connectivity matrix $W_{group} = \{w_{ij}\}$ with n nodes, we first calculate the full topological distance matrix D between all node pairs based on the Dijkstra's algorithm and using the inverse of the connection strength w_{ij} between nodes i and j as an initial topological distance. Subsequently, we derive the information measure I_{ij} between nodes i and j given by $1/D_{ij}$. I is then normalized by the maximum information measure. The NDI score of node m is then given as

$$NDI_m = mean(\{I_i\}_{i=1,..,m-1,m+1,...,n})$$
$$= mean(\{\sum_j I_{ij} - (I_{-m})_{ij}\}_{i=1,..,m-1,m+1,...,n}),$$

where $(I_{-m})_{ij}$ is the information measure of all nodes in the connectome from which node m has been removed. This analysis is then repeated for all nodes in the network, resulting in an nx1 dimensional feature vector of NDI scores for the network.

For each group connectome, we calculate its NDI scores. All nodes with an NDI of 0 are assigned to Tier 4. Using the natural-log-transformed NDI (excluding nodes with NDI = 0), we apply a Gaussian Mixture Model with 3 Gaussian distributions (GMM), where subnetwork assignments are based on the halfway point between the Gaussian centers, resulting in three additional Tiers. In total, the nodes in a network are differentiated into four Tiers (including the NDI = 0 Tier), where nodes within a Tier are "similar" with respect to their information measure.

2.4 Network Measures

We utilize the subnetworks defined on the group connectome to stratify the nodes in each subject's connectome. Subsequently, we characterize each subnetwork's topology by calculating two commonly used network measures describing different aspects of the connectome organization, namely transitivity (T), and global efficiency (E) [15].

2.5   Statistical Analysis

First, we compare the ranking of the regions in the brain, defined by NDI, between the different group connectomes. NDI assigns a measure of "importance" to each node in the connectome. According to this assignment, we can subsequently rank the regions in the brain. In order to compare different assignments from multiple group connectomes, we ultimately compare the resulting ranked lists. Here, we use the ranked-biased overlap (RBO) measure to estimate similarity of rankings, with higher weights for higher ranks [23]. In our setting this means that a variation in the order of important nodes is penalized more strongly, compared to the order among less important nodes. The closer the RBO value is to 1.0, the greater the agreement in node ranking between the three connectomes, with 1.0 representing complete similarity.

Subsequently, we calculate the network measures for each of the subnetworks defined on each group connectomes. This allows us to investigate topological differences between ASD patients and controls. We utilize the Mann-Whitney-Wilcox test and statistical significance was set at $p < 0.05$.

3   Results

The log-transformed NDI scores with the fitted GMM model are shown in Fig. 1 for each group connectome and for abs and pos weighting schemes. In case of retaining only negative weights, the population of calculated NDI value was so restricted, that no Gaussian fit was possible on neg matrices for all group connectomes. Therefore, for the remainder of this analysis, we restricted our analysis to abs and pos weightings only. In general, NDI scores are stable across group connectomes, with similar distributions and identified GMM centers within each threshold/weighting combination.

Table 2 summarizes the comparison of nodal ranking according to the NDI score. Firstly, for all combinations of threshold and weighting scheme, all three group connectomes resulted in differences of RBO values of less than 0.005. Secondly, RBO measures remained relatively consistent irrespective of threshold but differed more with weighting - meaning that edge thresholding has less effect on varying node rank across groups than weighting scheme. As node rankings were the same for all group connectomes for all threshold/weighting combination, we selected the Tier labels from the control connectome to stratify the nodes in each subject's network for the remainder of the analysis.

Figures 2 and 3 compare network topological measures in each of the subnetworks between ASD and controls, for absolute and positive weighting schemes, respectively. In the case of an edge threshold of 0.03 on the abs weighting, only one node had a value of NDI = 0, which meant that the network measures computed were ill-defined on this subnetwork (Tier 4). We observe significant differences in both T and E in Tier 1 in the case of using the absolute weighting for all thresholds, and in Tier 1 and Tier 2 in case of positive weighting, for a threshold of 0.01 and 0.05, respectively.

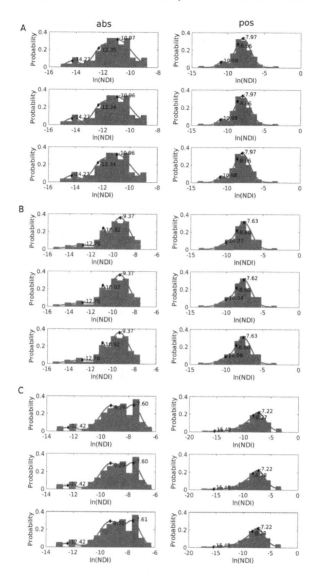

Fig. 1. Log-transformed NDI histograms for group connectomes based on absolute (abs; left column) and positive (pos; right column) edge weights. A, B, and C, correspond to edge thresholds of 0.01, 0.03, and 0.05, respectively. Each row of A, B, and C, corresponds (from top to bottom) to the cohort, patient-only, and healthy-only connectome. Centers of the three fitted Gaussians are indicated with a black diamond and the corresponding value is given to its right.

For absolute weights only Tier 1 showed significant differences. Regions which were consistently identified as Tier 1 regions across thresholds were the precentral gyrus, median cingulate and paracingulate gyri, cuneous, precuneous,

Table 2. Summary of RBO measures calculated on all group connectomes and for absolute (abs) and positive (pos) weights. Individual results of group connectomes are summarized by a single number, as they resulted in RBO differences of less than 0.005.

Threshold	Threshold	0.01		0.03		0.05	
	Weighting	abs	pos	abs	pos	abs	pos
0.01	abs	1.00	0.69	0.81	0.68	0.74	0.61
	pos	0.69	1.00	0.72	0.85	0.79	0.76
0.03	abs	0.81	0.72	1.00	0.68	0.81	0.62
	pos	0.68	0.85	0.68	1.00	0.73	0.78
0.05	abs	0.74	0.79	0.81	0.73	1.00	0.67
	pos	0.61	0.76	0.62	0.78	0.67	1.00

superior occipital gyrus, fusiform gyrus, and the supramarginal gyrus, all in the right hemisphere. Left hemisphere Tier 1 regions consisted of the orbital part of inferior frontal gyrus and medial superior frontal gyrus. Only the insula was identified bilaterally. For positive weights, we observed significant differences in Tier 1 and Tier 2 for a threshold of 0.01 and 0.05, respectively. Regions that were consistently identified in both tiers were the orbital part of middle frontal gyrus, median cingulate and paracingulate gyri, and caudate. In the left hemisphere, the regions consisted of the superior and inferior parietal gyrus, and the putamen. Additionally, the triangular part of inferior frontal gyrus was identified in both hemispheres.

4   Discussion

In this study, we showed that the NDI framework for defining subnetworks can be applied to resting-state functional connectomes and demonstrated its consistency regardless of the group connectome used. Additionally, we demonstrated that there are topological group differences in the subnetworks generated, when comparing subjects diagnosed with ASD and typically developing individuals.

Applying NDI for subnetwork definition worked for functional networks, utilizing an atlas with approximately half the number of regions, compared to the original study (which employed the Craddock200 atlas with 170 regions [19]). While it is possible that a reduction in the number of connectome regions can increase the variation in the GMM fitted Gaussian centers, we observed stable estimations for these centers for our three ABIDE group connectomes. In addition, we showed better agreement of nodal assignment, if weighting scheme is held constant while varying the threshold, compared to varying weighting scheme while holding the threshold constant. This further highlights the stability for the reliable estimation of subnetworks using NDI. Importantly, we showed that by stratifying the individual connectomes by subnetworks, we were able to find significant group differences between individuals with ASD and controls in regions

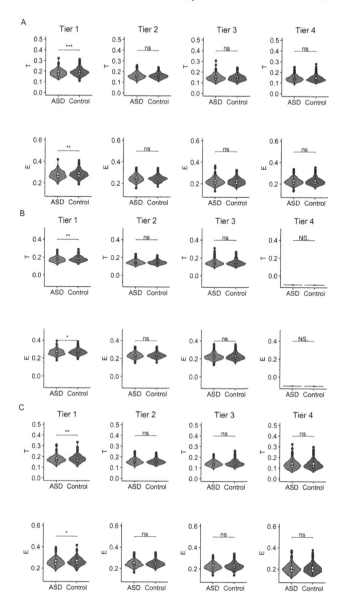

Fig. 2. Boxplot of topological network measures computed using absolute edge weights, for both ASD and Control groups from Tiers 1 to 4. A, B, and C correspond to edge thresholds of 0.01, 0.03, and 0.05, respectively. Statistical significance based on the Mann-Whitney-Wilcox test is indicated above each boxplot for transitivity (T) and efficiency (E). (ns: $p > 0.05$; *: $p < 0.05$; **: $p < 0.01$; ***: $p < 0.001$)

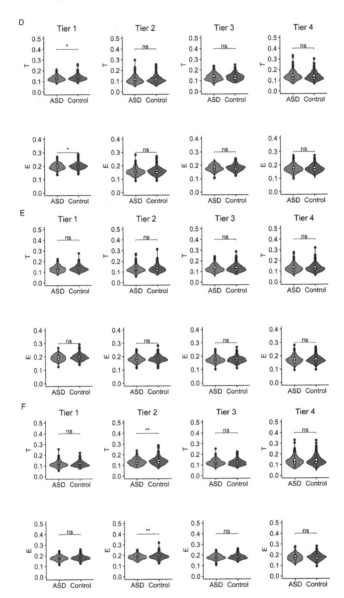

Fig. 3. Boxplot of topological network measures computed using positive edge weights, for both ASD and Control groups from Tiers 1 to 4. D, E, and F correspond to edge thresholds of 0.01, 0.03, and 0.05, respectively. Statistical significance based on the Mann-Whitney-Wilcox test is indicated above each boxplot for transitivity (T) and efficiency (E). (ns: $p > 0.05$; *: $p < 0.05$; **: $p < 0.01$; ***: $p < 0.001$)

belonging to Tier 1 or Tier 2, i.e. regions which are more important for efficient information transport. The only region consistently identified across thresholds and weighting schemes was the right median cingulate and paracingulate gyri, which was recently highlighted in ABIDE using a neural network approach [3]. Subsequent analyses may use this information, along with other regions identified from the appropriate weighting scheme, as priors when aiming to investigate specific regions.

There are general network analysis limitations, which may further affect our study. Although it is quite common to threshold functional connectivity matrices in order to reduce the effect of noise, there is no agreed upon method on how to define this threshold. Therefore, studies commonly investigate different thresholds with the aim to demonstrate consistency of results. Following this reasoning, we investigated a variety of methods of thresholding, specifically by using only the positive and only the negative edge weights, as well as the absolute edge weight, and by thresholding each at levels of 0.01, 0.03, and 0.05. While we show that the framework can still be utilized, except in the case of negative-only weights, there are many more thresholds which can be investigated. This will be the aim of future work. In our study, we analyze connectomes with 90 regions, whereas the original publication was able to utilize 170. While the appropriate atlas, or number of regions in the brain, remains an open question, a larger number of regions results in more data on which the three Gaussians can be estimated. In future work we aim to investigate agreement of NDI based Tier-assignment by utilizing multiple atlases, and mapping our results back to the brain template to identify spatial patterns of the regions in each Tier. In this work we estimated NDI subnetwork definition based on average group connectomes. However, it is possible to use the connectomes of each individual subjects to determine the subnetworks, which may help to further differentiate subtypes of diseases. While this is an interesting objective for future work, the primary aim here was to demonstrate that the NDI framework can be utilized in functional data and that it can identify group differences in case of disease.

In conclusion, we demonstrated that the NDI subnetwork framework can be applied to functional connectomes and produces stable results, when modifying the population from which the group connectome is generated (patients versus control). In addition, we show that these subnetwork definitions can be utilized to show group differences between individuals diagnosed with ASD and healthy controls, where those differences are mainly located in nodes/brain regions with highest importance.

Acknowledgments. This project has received funding from the European Union's Horizon 2020 research and innovation programme under the Marie Sklodowska-Curie grant agreement No. 753896 (MDS) and the American Heart Association and Children's Heart Foundation Postdoctoral Fellowship, 19POST34380005 (AWC).

References

1. Baio, J., et al.: Prevalence of autism spectrum disorder among children aged 8 years-autism and developmental disabilities monitoring network, 11 sites, United States, 2014. MMWR Surveill. Summ. **67**(6), 1 (2018)
2. Ball, G., et al.: Rich-club organization of the newborn human brain. Proc. Nat. Acad. Sci. **111**(20), 7456–7461 (2014)
3. Bi, X., Liu, Y., Jiang, Q., Shu, Q., Sun, Q., Dai, J.: The diagnosis of autism spectrum disorder based on the random neural network cluster. Front. Hum. Neurosci. **12**, 257 (2018)
4. Chung, A.W., Mannix, R., Feldman, H.A., Grant, P.E., Im, K.: Longitudinal structural connectomic and rich-club analysis in adolescent mTBI reveals persistent, distributed brain alterations acutely through to one year post-injury. arXiv:1909.08071 [q-bio.NC], pp. 1–22, September 2019
5. Chung, A.W., Pesce, E., Monti, R.P., Montana, G.: Classifying HCP task-fMRI networks using heat kernels. In: 2016 International Workshop on Pattern Recognition in NeuroImaging (PRNI), pp. 1–4. IEEE (2016)
6. Chung, A.W., et al.: Characterising brain network topologies: a dynamic analysis approach using heat kernels. Neuroimage **141**, 490–501 (2016)
7. Collin, G., Kahn, R.S., de Reus, M.A., Cahn, W., van den Heuvel, M.P.: Impaired rich club connectivity in unaffected siblings of schizophrenia patients. Schizophr. Bull. **40**(2), 438–448 (2013)
8. Craddock, C., et al.: The neuro bureau preprocessing initiative: open sharing of preprocessed neuroimaging data and derivatives. Front. Neuroinform. **7** (2013). https://doi.org/10.3389/conf.fninf.2013.09.00041
9. Di Martino, A., et al.: The autism brain imaging data exchange: towards a large-scale evaluation of the intrinsic brain architecture in autism. Mol. Psychiatry **19**(6), 659 (2014)
10. Grayson, D.S., et al.: Structural and functional rich club organization of the brain in children and adults. PLoS ONE **9**(2), e88297 (2014)
11. Hong, S.J., et al.: Atypical functional connectome hierarchy in autism. Nat. Commun. **10**(1), 1022 (2019)
12. Ktena, S.I., et al.: Brain connectivity measures improve modeling of functional outcome after acute ischemic stroke. Stroke, published online ahead of print 12, September (2019). https://doi.org/10.1161/STROKEAHA.119.025738
13. Müller, R.A., Fishman, I.: Brain connectivity and neuroimaging of social networks in autism. Trends in Cogn. Sci. **22**, 1103–1116 (2018)
14. Nielsen, J.A., et al.: Multisite functional connectivity MRI classification of autism: ABIDE results. Front. Hum. Neurosci. **7**, 599 (2013)
15. Rubinov, M., Sporns, O.: Complex network measures of brain connectivity: uses and interpretations. Neuroimage **52**(3), 1059–1069 (2010)
16. Rudie, J.D., et al.: Altered functional and structural brain network organization in autism. NeuroImage Clin. **2**, 79–94 (2013)
17. Schirmer, M.D.: Developing brain connectivity: effects of parcellation scale on network analysis in neonates. Ph.D. thesis, King's College London (2015)
18. Schirmer, M.D., Chung, A.W.: Structural subnetwork evolution across the lifespan: rich-club, feeder, seeder. In: Wu, G., Rekik, I., Schirmer, M.D., Chung, A.W., Munsell, B. (eds.) CNI 2018. LNCS, vol. 11083, pp. 136–145. Springer, Cham (2018). https://doi.org/10.1007/978-3-030-00755-3_15

19. Schirmer, M.D., Chung, A.W., Grant, P.E., Rost, N.S.: Network structural dependency in the human connectome across the life span. Netw. Neurosci. **3**, 792–806 (2018)
20. Schirmer, M.D., et al.: Rich-Club organization: an important determinant of functional outcome after acute ischemic stroke. Front. Neurol. **10**, 956 (2019). https://doi.org/10.3389/fneur.2019.00956
21. Van Den Heuvel, M.P., Sporns, O.: Rich-club organization of the human connectome. J. Neurosci. **31**(44), 15775–15786 (2011)
22. Varoquaux, G., Craddock, R.C.: Learning and comparing functional connectomes across subjects. NeuroImage **80**, 405–415 (2013)
23. Webber, W., Moffat, A., Zobel, J.: A similarity measure for indefinite rankings. ACM Trans. Inf. Syst. (TOIS) **28**(4), 20 (2010)
24. Zhao, T., et al.: Age-related changes in the topological organization of the white matter structural connectome across the human lifespan. Hum. Brain Mapp. **36**(10), 3777–3792 (2015)

Author Index

Printed in the United States
By Bookmasters